Guide to the Study of ANIMAL POPULATIONS

GUIDE TO THE STUDY OF ANIMAL POPULATIONS

James T. Tanner

THE UNIVERSITY OF TENNESSEE PRESS : KNOXVILLE

Library of Congress Cataloging in Publication Data

Tanner, James Taylor.
Guide to the study of animal populations.
Bibliography: p.
Includes index.
1. Animal populations. I. Title
QH352.T36 591.5'24 77-13630
ISBN 0-87049-235-7

Preface

An early measurement of a population is recorded in the Gospel according to Luke: about two thousand years ago Caesar Augustus ordered that all the individuals in the Roman Empire be counted. Although this was one of the first and simplest steps in the study of populations, several centuries passed before most nations of the world made regular censuses. More detailed analysis of population characteristics began as early as the third century when age at death was used to estimate the probabilities of death for persons of different ages, information necessary for the operation of life insurance and annuities.

The statistical study of human populations is now the well-established science known as demography. Its basic data come from censuses and from the official registration of births, deaths, and marriages.

The concept that other animals live in groups having some of the same properties as those of humans existed at least by the twelfth century (Leopold 1933). These groups easily came to be called "populations," even though the word is derived from one for "people." By definition, a population is a group of individuals of one species occupying a geographic or similarly defined region. As a group it has certain characteristics such as the density of individuals, a birth rate, and so on. It is always changing because of births, deaths, and migrations; to adapt the old saying about a river, one never sees the same population twice. Therefore it is well to enlarge the above definition to say that a population consists of the individuals of one species present in a defined area at a certain time.

The study of animal populations has taken two directions. One is the theoretical or mathematical study of population dynamics, beginning with the works of Verhulst (1838), Lotka (1925), and Volterra (1926). A second is the study of the means of manipulating populations for practical or economic goals. This is illustrated in Leopold's classic book on game management (1933), in which he makes the subject of population characteristics the central theme of the book. Other examples of the management of animal populations are in fisheries, the control of pest insects, and the preservation of rare species. Both the theoretical and practical aspects depend upon an understanding of population characteristics and upon the ability to measure these.

My purpose here is to describe the characteristics of animal populations, the methods of their measurement, and their interactions. These characteristics can be classed into those describing the population as it exists at some instant (e.g., density and composition) and those measuring the ways in which it is changing (e.g., birth and death rates). An example of an interaction is the effect of age composition on the birth rate. I have intended to review the important aspects of the subject and to describe the research methods most useful in animal population studies, emphasizing those methods most amenable to statistical analysis.

Whenever reasonable, I have used the concepts and terms of demography, a better-established science than that of animal populations. Good introductions to demography are by Keyfitz (1968) and Spiegelman (1968). This approach is limited, however, because data on animal populations cannot be obtained as in demography; most animals will not stand still and be counted, and individuals have no official identity, no birth certificate.

A complete description of each topic, including mathematical proofs and steps for each calculation, would be unbearably long, and so I have sought a compromise. Mathematical proofs are rarely given; instead I have usually cited a published exposition of a proof or have given an indication of the course of the proof so that the reader can work it out for himself. I assume that the reader has had at least basic courses in calculus and statistics. For the most frequently used measurements and analyses, the procedures are outlined to facilitate setting up the attack on a problem and worked examples are given. I discovered that it would not be practical to refer to all the methods of population analysis which have been described in the literature, so I confined the text to those which have been tried and found to work. Nor was it practical to prepare a compendium of field technics, since each is usually limited to a particular animal group, but many citations are given to descriptions of technics.

I should say what this book is not. It is not a discussion of population dynamics, of species interactions, of stability or cycles. It is not a review of population modeling or simulation. It does, however, aim to explain the attributes of populations upon which the above subjects are based, and it provides a brief introduction and references to each of these. Excluded entirely is the really different subject of population genetics.

Contents

Figures & Tables

FIGURES

TABLES

Symbols

Symbols are used as consistently as possible throughout the following chapters. To do this, it was occasionally necessary to change from those commonly used in the literature; in these cases the customary symbols are mentioned along with those chosen. Capital letters are generally used for quantities which are counted or measured (e.g., N for the number of individuals). Lower-case letters symbolize rates, and so forth, which are obtained from ratios or proportions; an example is b for the birth rate. A subscript (usually lower case) identifies a number or rate at a particular time, or of some part of the population, etc.; N_t is the number at time t, N_f the number of females.

The following symbols are the same as those commonly used in mathematics or statistics:

- $\overline{N}$ (read "N bar"): the mean value of N, the arithmetic average;
- N (read "$\hat{N}$ hat"): the estimated value of N;
- μ (mu): the true or real mean value, usually unknown;
- σ^2 ("sigma squared"): the true variance (e.g., of the population density);
- S^2: the variance calculated from the samples;
- S: the standard deviation of the samples;
- Δ (read "delta"): a change in the value of some quantity, an increment or a decrement;
- e: the base of natural logarithms (equals approximately 2.7183);
- ln: the natural logarithm of a number (e.g., if $e^a = x$, $ln\ x = a$;
- f: indicates some mathematical function of a variable (e.g., $N = f(t)$ means that N is some unspecified function of t);
- $P(\ )$: the probability of (e.g., $P(x < b) = .05$ means that the probability that x is less than b is the given figure);
- Σ: the sum of—with the following variations—

  $\Sigma_{i=1}^{k} N_i = N_1 + N_2 + N_3 + \ldots + N_k$;

$\Sigma_i N_{ij} = N_{.j}$
(for one value of j the N_i are summed over the range of i);
ΣN
(with no subscript: the sum of all values of N in the sample);
Π : the multiple product of
$\Pi_{i=1}^{k} s_i = s_1 \times s_2 \times s_3 \times \ldots \times s_k$;
! (read "factorial): for example, $4! = 4 \times 3 \times 2 \times 1$, and likewise for any other integer;
$\neq$: is not equal to.

There is no standard set of symbols used for population characteristics, excepting a few which appear quite consistently. So I have used those which have worked well during several years of teaching the subject. They are listed below with the page numbers on which each is introduced and defined.

b: birth rate, p. 119
b_a (or b_{fa}): age-specific maternity rate, p. 121
C_a: proportion of the population between ages a and $a + 1$, p. 135
d: death rate, pp. 70–71
$\bar{e}_0$: expectation of life, average longevity, p. 82
g: growth ratio, p. 127
l_a: probability of surviving to age a, p. 83
L_a: survivors of a cohort to age a, p. 81
m: mortality rate, pp. 70–71
r: population growth rate, p. 125
r_m : intrinsic rate of population growth, p. 154
R_0: net reproductive rate or generation ratio, p. 121
s: survival rate, p. 71
$\bar{T}$: mean length of a generation, pp. 149–50
T_c : cohort generation time, p. 149
T_D : doubling time for a population, p. 128

1 Density and Dispersion Measured by Sampling

Questions frequently asked about a population are: How big is it? Are the individuals crowded or are they thinly scattered? Are they spaced evenly throughout the area inhabited by the population or are they clumped into herds or similar groups? The first question is rarely explicitly answered for animal populations: however, it is implied in the answer to the second one, which deals with population density, a measurement determined by the number of individuals and the area inhabited. The third question, about the pattern in which the individuals occur, is answered by measuring the population dispersion. Both of these characteristics, density and dispersion, are important in ecology and in management for many reasons, and their measurement is the object of many population studies.

Density and dispersion need to be studied together for the following reasons: dispersion refers to small scale variations in density; the measurement of one is affected by variations in the other; and both of them can be measured together by sampling methods. This chapter describes sampling methods of the usual kind (e.g., one in which the individuals in a sample plot can be counted). Chapter 2 describes variations of sampling which have been developed for animals, and chapters 3 and 4 discuss the estimation of numbers by, respectively, the mark-recapture method and changes in sex or age ratios.

1.1 DEFINITIONS

Density is the number of individuals per unit area occupied by the population (e.g., the number per square meter or per hectare). For some animals, such as planktonic species, density is better defined as the number per unit volume. The inverse of density (e.g., hectares per individual) has sometimes been used as a measure when populations are scarce to avoid reporting a fraction of an individual per unit area, but it is preferable to use the direct measure and if necessary to use units like 10 hectares or 100 square meters to allow reporting in convenient figures. An example would be 5 birds per 10 hectares (not .5 bird / hectare or 2 hectares / bird).

General density is measured with respect to all of the area containing the population. *Specific density* is measured with respect to some part of the area, such as that of the particular or preferred habitat of the species. This has also been called the "ecological density" and is similar to "population intensity," which is the number per unit of habitat such as a leaf or shoot (Southwood 1966).

Density can be calculated from a *census*, a count of all the individuals in the population, the count to be divided by the area occupied. Censuses are performed on human populations, where each individual can be recognized and counted. They are, however, rarely possible for either animal or plant populations.

Dispersion refers to the pattern of individuals in space—to the way in which density varies in small areas of the habitat. There are three major patterns of dispersion (see Fig. 1.1): *random*—when there is an equal probability of any point in space being occupied, and the location of one organism does not affect the position of another; *aggregated*—when the individuals are clumped, often called a "contagious" pattern; *regular*—when the individuals are spaced more or less regularly, sometimes called a "uniform" pattern.

Dispersion of a population may result from two separate or inter-

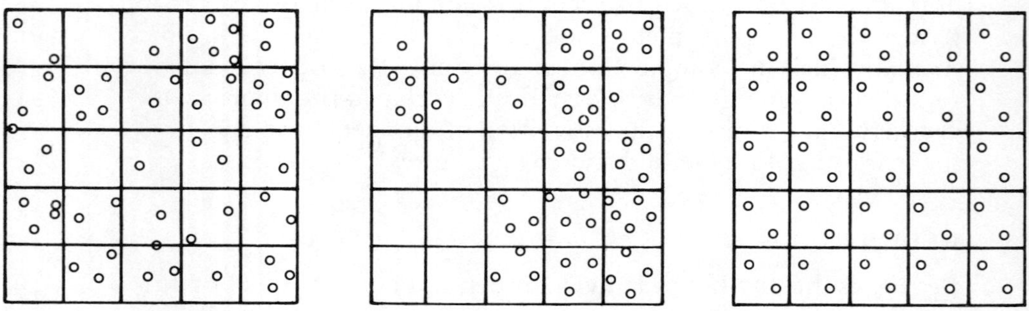

A. Random B. Aggregated C. Regular

Figure 1.1. Patterns of dispersion. The grid serves to show the effects of the different dispersions on the variance of the sample counts.

acting causes. One is variation in the distribution of environmental factors (e.g., the dispersion of suitable attachment or nesting sites). The second is intraspecific behavior, such as an aggregated dispersion resulting from social behavior, or a regular dispersion caused by competition for space.

1.2 SIMPLE RANDOM SAMPLING

"The estimation of populations is both a science and an art," wrote J. G. Skellam (1962). It is a science in that knowledge is acquired about a specified population by an inductive process which can give reproducible results. It is an art in that the actual method of counting or estimating numbers depends upon knowledge of the species and the skill of the worker. Knowledge of the species is also required because it is usually possible to examine only a part or a sample of the population, and it is important that this sample represent the population under study. The estimation of population density by sampling is the subject of this section.

Density may be measured by taking a sample of the space inhabited by the population and then counting the individuals in each sampling unit. Usually there are several sampling units, often called plots or quadrats, in one sample. The results may be simply stated as the estimated density equaling (total individuals counted)/(total area sampled). However, if certain sampling procedures are followed, more information than this can be obtained about both the population and the usefulness of the estimate. For example, the estimate can be qualified as to its accuracy and its precision. The first of these qualifications refers to the difference between the estimated density and the true population density; since the latter is probably not known, accuracy cannot really be determined, but it may be evaluated by the calculation of a confidence interval or by analyzing the bias—a persistent error in one direction—of the sampling method. Precision is a measure of the scatter or the deviation of the counts from the sample mean and is measured by the sample standard deviation, which becomes smaller with increasing precision.

Three decisions must usually be made in planning a sampling program so as to increase both the accuracy and the precision of the results: 1) the method to be used in counting or estimating the number in a sampling unit; 2) the procedure for locating or selecting sampling units; and 3) the related questions of the size and shape of each sampling unit and the number of units to be taken.

Methods of Collecting and Counting

Below are listed some reviews or summaries of methods used in sampling different groups of organisms:

Planktonic organisms: Edmondson and Winberg 1971;
Benthic organisms: Edmondson and Winberg 1971, Elliott 1971;
Soil-inhabiting animals: Macfadyen 1962, Murphy 1962, Southwood 1966;
Insects: Lamotte 1969, Southwood 1966;
Fish: Lagler 1971;
Birds: Giles 1971, Hall 1964, Lamotte 1969, Kendeigh 1944;
Mammals: Bergerud 1968, Giles 1971, Lamotte 1969;
Terrestrial plants: Grieg-Smith 1964.

There are numerous descriptions in the literature of methods adapted to sampling a particular species. A bibliography on sampling (Schultz *et al*, 1976, p. 281–98) lists about a hundred titles on sampling of particular species and others on general theory and methods.

Location of Samples

A requirement of most statistical tests applied to sampling data is that each point in the area being sampled have the same probability of being included as every other point. If this condition is met, the sampling is called "random." One means of achieving a random sample is to divide the entire area into small plots by a numbered grid and then to select some plots for the sample by using a table of random numbers. This procedure, when applied over the entire area inhabited by the population under study, is called "unrestricted" or "simple random sampling."

Shape, Size, and Number of Sampling Units

Units may have a geometric shape such as a circle on the ground or they may be units of the habitat such as a plant or the branch of a tree. The choice must be made on the basis of practicality with the goal being to obtain accurate data with the least time and effort. Most sampling is done by selecting plots on a surface such as the ground, so the major advantage of each of the three simplest shapes will be mentioned. A circle has the least boundary for the enclosed area so will produce the fewest problems with counting organisms that are on or near the boundary. A square is frequently the easiest to lay out and measure, particularly if a grid is being used to select plots randomly. An oblong may include more variation in the habitat, thus making the sample plots more like each other and producing counts with smaller variance. There is no one best shape.

Consideration of the size and number of plots raises these questions: Is it better to have many small plots or a few large ones? What is needed for statistically useful results? To answer these it will be helpful to begin with a discussion of the expected distribution of the number of individuals per plot. An understanding of this distribution will also be useful in studying the dispersion of organisms.

If the individuals are dispersed randomly throughout the area being sampled, the number of individuals per plot has the Poisson distribution. Demonstrations of this may be found in Grieg-Smith (1964, p. 12) and in Pielou (1969, p. 81). The conclusion may be stated as follows: Given that the individuals are distributed randomly with a true mean of μ individuals per plot, the probability that any one plot will contain N individuals is

$$P(N) = \frac{\mu^N}{e^{\mu} N!} . \tag{1.1}$$

In using equation (1.1) to calculate the probability of a plot containing exactly N individuals, μ is replaced with $\overline{N}$, the measured sample mean. The powers of e and the values of factorials can be found in mathematical tables. An illustration of the calculation is given in Example 1.1.

Ex. 1.1. Use of the Poisson distribution to calculate the probability of number per sampling unit. The numbers of individuals in 50 units were counted, and the mean was 9. If the dispersion is random, what is the probability of a sampling unit containing exactly 5 individuals? 9 individuals? 15 individuals? From tables, $e^9 = 8103$.

$$P(5) = \frac{9^5}{8103 \times 5!} = .06$$

$$P(9) = \frac{9^9}{8103 \times 9!} = .13$$

$$P(15) = \frac{9^{15}}{8103 \times 15!} = .019$$

Compare these probabilities with those plotted in figure 1.2 for $u = 9$ and $N = 5, 9, 15$. The expected number of units with 5, 9 or 15 individuals is respectively: $.06 \times 50 = 3$; $.13 \times 50 = 6.5$; $.019 \times 50 = .95$.

Figure 1.2 shows how the Poisson distribution changes as the mean increases. It is very unsymmetrical when the mean is 3 or less, but with the mean equaling 9 the distribution is quite symmetrical, and it becomes more so with larger values of the mean. Because a Poisson distribution approaching symmetry resembles a normal distribution, the sample mean $\overline{N}$ will be almost normally distributed around the true mean μ with standard error $S_{\overline{N}}$ equal to $S/\sqrt{Q}$, where Q is the number of sampling units used in calculating $\overline{N}$. This leads to the conclusion that if $\overline{N}$ is large enough for the Poisson distribution to be nearly symmetrical, $\overline{N}$ and $S_{\overline{N}}$ may be used with tables of the normal distribution to estimate confidence intervals, etc., subject to the condition described in the next paragraph.

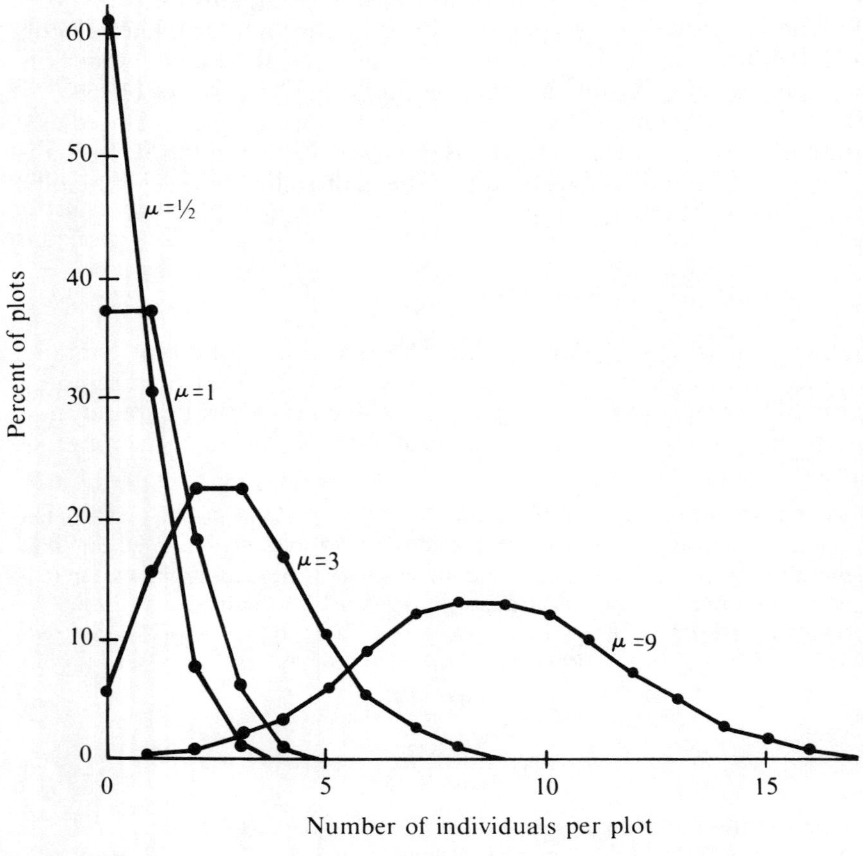

Figure 1.2. Poisson distributions for means of ½, 1, 3, and 9.

A peculiarity of the Poisson distribution is that the variance and the mean are equal and are therefore not independent of each other. For statistical tests requiring that the mean and variance be independent or that the variances of two or more groups be equal, the Poisson variable can be transformed to $x = \sqrt{N}$; x will have a stable variance, and the transformation will probably also reduce the asymmetry.

To return to the problem of sample size: Since $\overline{N}$ will increase with the size of the plots or sampling units, these should be large enough for the expected Poisson distribution to be reasonably symmetrical; this can easily be tested by using the data to plot a frequency distribution such as that in figure 1.2. There is little to be gained by increasing the size of each sample beyond this point; instead, more information will be gained by increasing the number of sampling units. More information on determining the size of the sample is presented later in this section.

Estimates from Simple Random Sampling

The data obtained from this sampling procedure can be used to estimate population density and number and the precision of the estimate by the calculations set out below. More information on these and the proofs of the equations may be found in a text on sampling procedures such as that by Cochran (1963). The following symbols are used:

Q: the number of units in the sample;
$\overline{N}$: the mean number of individuals per unit;
S: the sample standard deviation;
f: the sampling fraction, equals Q/(total number of sampling units possible), or the fraction of the total population included in the sample.

The sampling fraction is used in what is called the finite population correction, which is $(1 - f)$. If the sampling fraction is small (e.g., 10 percent or less, as is usually the case in samples of animal populations), the finite population correction can be omitted; that is, it can be set equal to 1.

The population density is estimated by $\overline{N}$, and the precision of this estimate by the standard error of $\overline{N}$,

$$S_{\overline{N}} = S\sqrt{(1-f)/Q}. \tag{1.2}$$

To estimate the entire population from $\overline{N}$ we need to know the total number of sampling units possible, which equals Q/f. Then the population estimate $\hat{N} = \overline{N}Q/f$ and its standard error $S_{\hat{N}} = S_{\overline{N}}Q/f$. Example 1.2 illustrates the above calculations.

Ex. 1.2. Estimates from one random sample. Twenty large plots were randomly spaced in an area occupied by a small population; these units covered 10 percent of the whole area, so $(1 - f) = .9$. The sample mean $\overline{N} = 212$, and the sample standard deviation $S = 15$. From these data the following estimates can be made: The standard error $S_{\overline{N}} = 15\sqrt{.9/20} = 3.2$. The estimated total population is $\hat{N} = 212 \times 20/.1 = 42{,}400$. The standard error of $\hat{N}$, $S_{\hat{N}} = 3.2 \times 20/.1 = 640$. A 95 percent confidence interval for $\hat{N}$ can be estimated by assuming that $\overline{N}$ is normally distributed and so 95 percent of all estimates will lie within the interval of the true number ± 2 $S_{\hat{N}}$. The 95 percent confidence interval = 42,400 ± (2 × 640), or 41,120 to 43,680.

If it is desired to attain a certain precision of the estimate (i.e., a standard error of a certain size), and if the sample standard deviation is known, these can be substituted into equation (1.2) which can then be solved for Q, the required number of sampling units. To give a simple example, assume that f is so small it can be neglected and that the desired precision is 10 percent of $\bar{N}$; then $Q = (S/.1\bar{N})^2$.

Sample Size Required for Accuracy

The accuracy of any estimate obtained from a sample, assuming that the sampling method is unbiased, will depend upon the sample variance and the sample size. The variance for a given method is largely beyond the control of the research worker, but he can determine within reason the number of sampling units to be taken, or the answer to this question: How many sampling units will be needed to attain a certain accuracy of the estimate? There are two aspects to this accuracy: 1) the desired margin of error, plus or minus either some fraction of the mean or a definite amount irrespective of the mean; and 2) the lowest probability of being within that margin of error (or conversely, the highest acceptable probability of being in error). These aspects can be combined into a statement such as the following: To be able to estimate $\bar{N}$ so that with a probability of at least 95 percent the true mean μ will lie within 10 percent of $\bar{N}$, at least Q sampling units will be required. Such an estimate can of course be made only if a preliminary estimate of the mean and standard deviation is available, perhaps from a pilot study.

The following steps are used for calculating the number of samples required for a certain accuracy:

1) Obtain a preliminary estimate of $\bar{N}$ and the standard deviation S;
2) Decide on the allowable error E (e.g., 10 percent of $\bar{N}$);
3) Decide on the desired probability, and from a table of the standard normal distribution determine the value of $\pm t$ that will not be exceeded with the desired probability; for example, $t = 1.96$ for the probability of 95 percent;
4) The minimum number of units required to obtain the desired accuracy is

$$Q = \left(\frac{tS}{E}\right)^2 \qquad (1.3)$$

$$= \left(\frac{1.96\ S}{0.1\ \bar{N}}\right)^2$$

using the figures given for illustration.

Of course this is not a guaranteed answer. After taking a sample of size Q the research worker may find that his preliminary estimates of $\bar{N}$ and S were wrong, and he may need to take a larger sample; or, on the other hand, he may have taken one unnecessarily large. This is another illustration of Skellam's (1962) statement that "the estimation of populations is both a science and an art."

Ex. 1.3. Prediction of the number of samples required. A trial sampling of the number of aphids on each of 20 plants yielded a mean of $\bar{N}$ = 20 with a standard deviation S = 12; the distribution of insects per plant was reasonably symmetrical. The sampling program is to be continued with the goal of estimating $\bar{N}$ with an allowable error of $E = \pm 3$ with a probability of 5 percent of exceeding this error. The estimated number of samples for obtaining this accuracy is equation (1.3)

$$Q = \left(\frac{1.96 \times 12}{3}\right)^2 = 62.$$

Further discussion of the size of the sample is postponed until the end of the next section.

Nonparametric Tests Applied to Sampling Data

Some of the preceding discussion has described the desirability of sampling so that the data are symmetrically distributed to allow the use of statistical tests based on the normal distribution. However, Goodall (1970) has pointed out that because sampling distributions are usually asymmetrical and sample variances from natural populations are characteristically large, efforts to obtain precise measurements with a normal distribution may therefore be wasted. He suggested that it might be more profitable to use quick sampling methods that will yield abundant data for nonparametric tests, which generally require more samples for significant results. In other words, it may be better to sacrifice precision in order to gain a large mass of data which can be used in less restrictive tests. Nonparametric tests have infrequently been applied to ecological data, so two such tests are briefly introduced in the following paragraph.

The median test is easy to use for testing the null hypothesis that two or more populations have equal medians. The basic data are the number of individuals in each sampling unit. The number of units from each of the two or more populations can be unequal. It is not a powerful test, but it is conservative in the sense that if the null hypothesis is rejected it is done with confidence. The Mann-Whitney, or "U," test is much more powerful for comparing two populations, but it cannot be applied easily or safely if there are many units from the two populations with an equal number of individuals. Such ties can be avoided by taking larger samples; if this is done, the investigator has the choice of using this nonparametric test or one based on the normal distribution. The Mann-Whitney test is almost as powerful as the parametric *t*-test and is not restricted by as many conditions, so in that respect it would be the better. The necessity for large samples is a disadvantage, however, and in general it appears better to use "quick and dirty" sampling and the median test with its few restrictions. Nonparametric statistical methods are described in texts by Siegel (1956) and Conover (1971). Applications of some nonparametric tests to numbers of benthic invertebrates are described by Elliott (1971).

1.3 RESTRICTED AND STRATIFIED SAMPLING METHODS

The previous section dealt with the methods wherein the sampling units are located randomly throughout the area occupied by the population under study. In other methods the units are selected randomly within restricted blocks or strata for purposes of increasing precision or decreasing costs. If the selection is properly done, the necessary condition of randomness is maintained.

Block Random Sampling

The accuracy of an estimate may be increased by dividing the area inhabited by the population into a number of blocks of equal size and locating within each block an equal number of randomly placed sampling units (see figure 1.3). This will insure that each point has the same probability of being included, and accuracy may be increased

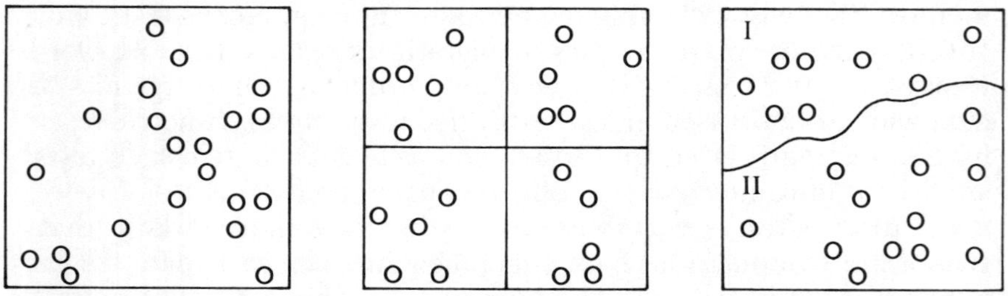

A. Simple random sampling. B. Block random sampling. C. Stratified random sampling.

Figure 1.3. Twenty sample plots selected by three different methods. In each example the horizontal and vertical coordinates of the samples were determined from a table of random numbers. (A) Twenty plots are placed randomly over the entire area. (B) Five plots are placed randomly in each of four blocks. (C) Eight plots are placed randomly in habitat I (40 percent of the area), and twelve plots are placed randomly in habitat II.

because all or most of the sample will not by chance come from one section having different characteristics from the rest of the area. The calculations for this method are the same as those for simple random sampling. Grieg-Smith (1964) discusses block random sampling as applied to plant populations.

Two-Stage Sampling

This method resembles the preceding in that the area is divided into blocks, but then some of the blocks are randomly chosen and only within these are sampling units randomly located. This procedure may be used to reduce the cost or work of sampling. Calculations of the estimates are different from those given above, because of the two random processes; these may be found in texts on sampling (see Cochran 1963, chap. 10).

Two-Phase Sampling

This method is often useful when a population of animals inhabits small, localized habitats, such as an insect occurring on one species of plant which is scattered throughout an area. In phase 1 the density of the habitats is estimated by sampling, and in phase 2 some of these are chosen as sampling units to estimate the density of the population within the habitat; from these an estimate of population density or size may be made. The procedure is described in texts on sampling (see Cochran 1963, chap. 12). Pielou (1974) illustrates two-phase sampling with an example of insects inhabiting flower heads.

Stratified Random Sampling

This is the most sophisticated and efficient method of sampling. The entire area is divided into sub-areas, the strata, each of which is a uniform habitat for the species being studied; this of course requires prior knowledge about the area and the habitat requirements of the species. Sampling units are chosen randomly within each stratum. The number of units in each stratum may vary; the simplest and usually the most efficient procedure is to take a number proportional to the area of the stratum (see figure 1.3). The advantage of stratified sampling is that estimates may be obtained separately for each habitat; these may then be combined into an estimate for the entire area which has greater accuracy or which was obtained with less expense than by other methods.

The preceding discussion of stratified sampling assumed that the existence of strata—areas or habitats in which the population density differed—was already recognized. If this is not so, a preliminary sampling program aimed at detecting the presence or absence of strata is desirable. The general procedure can be described with the aid of a hypothetical example, beginning with a situation in which the habitat can be identified into four, or two by two, overlapping classes. For example, these might be pine versus deciduous forest and old growth versus a young stand of each, or two depths of water each with sand or mud bottoms. Random samples are made in each of the four habitats and the number of animals in each sampling unit counted. The data can be tested by analysis of variance for significant differences in density. To make the explanation simpler, let us consider the first illustration, two types of forest, each of two ages. The null hypothesis is that the density is the same in all four habitats. If the analysis of variance accepts this hypothesis, there is no basis for dividing the area into strata. The null hypothesis may be rejected on the basis of a significant difference only between the forest types, in which case these become strata, or the two ages of trees similarly may be identified as strata. Or there may be significant differences in both directions showing the presence of four strata. This test is illustrated in example 1.4.

Watt (1968: 190–97) uses an example of a three-way classification (samples from different trees, from different levels above ground, and from different exposures) to describe the uses of analysis of variance in planning a stratified sampling program.

The equations for estimation from stratified random sampling are given below. The correction for the sampling fraction is omitted for simplification and because it is seldom needed in samples from animal populations. This correction is included in the equations given by Cochran (1963, chap. 3), who also gives the proofs of the equations and several corollaries. In the following, the subscript h indicates a stratum; Q_h, $\bar{N}_h$, and S_h are, respectively, the number of sampling units, the sample mean, and the sample standard deviation from stratum h. Additional symbols are:

W_h: weight of estimate from stratum h, equals proportion of stratum h to the entire area.

$Q.$: sum of all Q_h.

Q_T: total number of sampling units possible from entire area; e.g., if the units are plots of equal size, Q_T = (total area)/(area of 1 plot).

The weighted population mean over all strata is

$$\bar{N}. = \sum_h W_h \bar{N}_h. \tag{1.4}$$

The standard error of this mean is

$$S_{\overline{N}} = \sqrt{\sum_h (W_h S_h)^2 / Q_h}. \qquad (1.5)$$

The population total, over all strata, is

$$\hat{N} = Q_T \overline{N}.,$$

and its standard deviation is

$$S_{\hat{N}} = Q_T S_{\overline{N}}.$$

Ex. 1.4. A stratified random sample. The study area contained two forest types, type 1 and type 2, and in each were stands of two ages, *a* and *b*. Twenty plots were randomly located in each of the four combinations of type and age, and the numbers of a species of centipede were counted in each plot. To test for stratification, the data were subjected to a two-way analysis of variance, and the results were as follows:

	Sum of Squares	Degree of Freedom	Mean Square	F	
Type	520.2	1	520.2	59.79	($P<.01$)
Age	20.0	1	20.0	2.30	($P>.05$)
Interaction	3.2	1	3.2	-	
Residual	661.2	76	8.7	-	
Total	1204.6	79			

Since there was a significant difference only between types, the ages within each type were combined to make two strata: type 1, which covered 75 percent of the area, and type 2, covering 25 percent. The sample data are

$$Q_1 = 40, \quad \overline{N}_1 = 17.0, \quad S_1 = 3.08, \quad W_1 = .75,$$
$$Q_2 = 40, \quad \overline{N}_2 = 11.9, \quad S_2 = 3.03, \quad W_2 = .25.$$

The estimates made using equations (1.4) and (1.5) are

$$\overline{N}. = (.75 \times 17.0) + (.25 \times 11.9) = 15.7$$

and

$$S_{\overline{N}.} = \sqrt{(.75 \times 3.08)^2 / 40 + (.25 \times 3.03)^2 / 40} = 0.38.$$

The increased precision obtained from stratified sampling can be seen by comparing S_1 and S_2 with the standard deviation of all eighty plots; from the above table the latter is $S = \sqrt{1204.6/79} = 3.90$.

In this section and in the previous one we have discussed the related subjects of precision or accuracy and the number and selection of sampling units. Little was said about the costs, in time and money, of obtaining a sample. All of these subjects can be brought into perspective by considering the goals of a sampling program, which is essentially a plan for obtaining the most accurate estimates for the least cost. This ideal goal is obviously unattainable, so the preparation of a sampling program requires compromise and judgment. The main factors which need to be considered are included in the formula for the standard error, $S_{\overline{N}} = S/\sqrt{Q}$ (not including a finite population correction). Precision and accuracy are increased by reducing the standard error. One way to do this is to increase the number of sampling units, but the gain in accuracy is a function of the square root, while the increase in cost is usually a direct function of the number of units. The benefit therefore cannot be expected to be proportional to the increased cost. A decrease in the sample standard deviation, however, will directly reduce the standard error and increase precision. Whether or not this can be accomplished without costing too much will depend primarily on the skill and ingenuity of the research worker. One method of decreasing the sample standard deviation is that described in this section, the use of stratified sampling.

When the details of a sampling program are planned, information will be needed on the expected cost per sampling unit, the probable size of the sample standard deviation, and variations in the habitat which are likely to affect the standard deviation. This information may be obtained by a pilot sampling program. How all of this may be put together to increase sampling efficiency has only been outlined in these two paragraphs; details may be found in texts on sampling methods.

1.4 MEASURING POPULATION DISPERSION FROM SAMPLING DATA

The three basic kinds of dispersion—random, regular, and aggregated—have already been defined and are illustrated by figure 1.1. The dispersion of a particular population can be tested with sampling data, most easily by the ratio of the variance to the mean of the number per plot. Samples from a randomly dispersed population are expected to have the variance equal to the mean, or a variance : mean ratio of 1, since a random dispersion fits the Poisson distribution. If the pattern of a population were perfectly uniform, all plots would have about the same number of individuals (assuming that the plots are large enough so that none are empty) and the variance would be near zero; the variance : mean ratio would be less than 1. On the other hand, an extremely aggregated population would produce many counts of no or few individuals as well as many counts of high numbers, resulting in a large variance and a ratio greater than 1. Figure 1.4A compares the frequency distribution from a regularly dispersed population (vertical bars) with what would be expected in a randomly dispersed population having the same mean density (solid line,

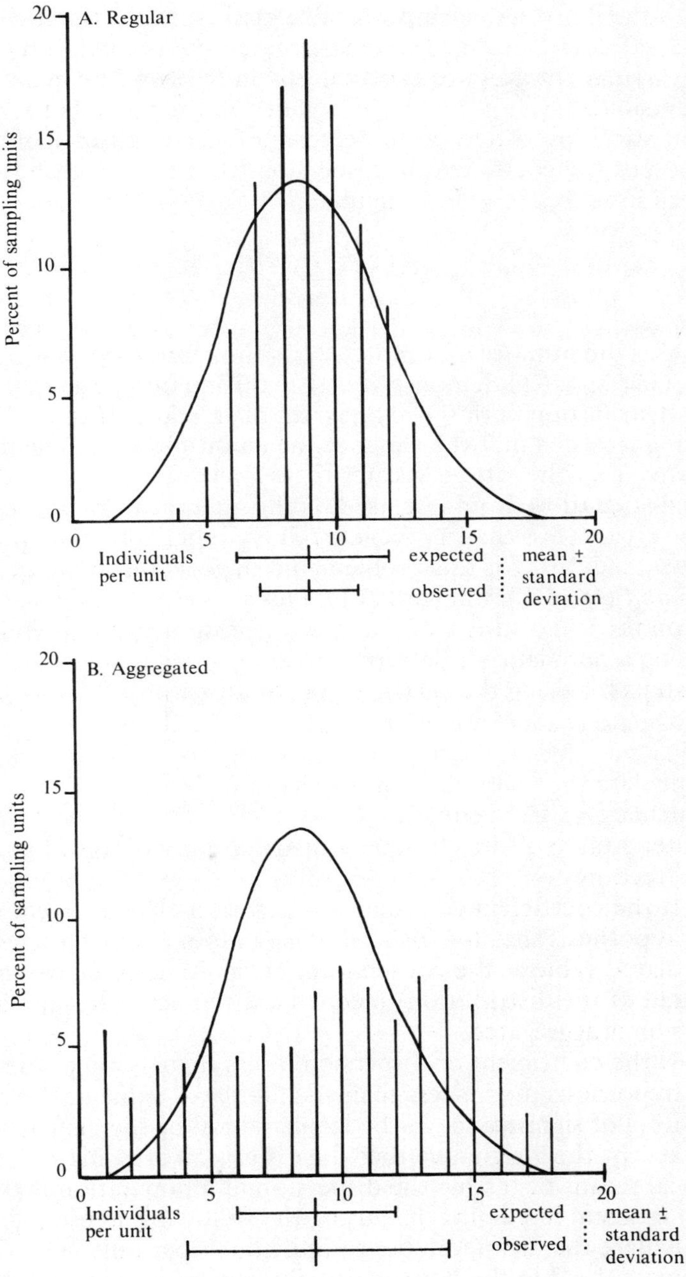

Figure 1.4. Observed distribution of numbers per sampling unit (vertical bars) compared with expected Poisson distribution (continuous curve). (A) from a regularly dispersed population; (B) from an aggregated population. Both distributions have a mean of 9. The mean ± the standard deviation are indicated by the horizontal bars at the bottom of each figure.

which is the Poisson distribution calculated from the observed $\overline{N}$); figure 1.4B does the same for an aggregated population.

The variance : mean ratio is often called the "coefficient (or index) of dispersion."

A statistical test is needed to determine the significance of any deviation of the coefficient of dispersion from the value of 1. The first step in such a test is to calculate

$$\chi^2 = \frac{\Sigma_{i=1}^{Q}(N_i - \overline{N})^2}{\overline{N}} = \frac{QS^2}{\overline{N}}, \qquad (1.6)$$

where Q is the number of samples, N_i is the number in sample i. Pielou (1969, p. 91) demonstrates that this quantity has the chi-square distribution with $(Q - 1)$ degrees of freedom. If the null hypothesis being tested is that the variance : mean ratio equals 1 against the alternative that the ratio is greater than 1, the right tail of the chi-square distribution should be used. If the alternative hypothesis is that the ratio is less than 1, the left tail is proper. The chi-square test is not an exact test for this problem and in general should not be used if $\overline{N}<5$ *and* $Q<15$ (Armitage 1971). Pielou (1969) adds some comments on the use of the coefficient of dispersion as a sample statistic describing a population's pattern.

The steps for using the variance : mean ratio are:

1. Calculate the mean number per unit, $\overline{N}$, and the sample variance, S^2.
2. Calculate the coefficient of dispersion, $S^2/\overline{N}$.
3. Calculate χ^2 from equation (1.6).
4. Enter a table of the chi-square distribution with $(Q - 1)$ degrees of freedom.
   a. If the coefficient of dispersion is greater than 1, reject the null hypothesis that the dispersion is random if χ^2 exceeds the tabulated value at the chosen level of significance in the right-hand tail of the distribution; accept the alternative that the dispersion is aggregated.
   b. If the coefficient of dispersion is less than 1, reject the null hypothesis if χ^2 is less than the tabulated value at the chosen level of significance in the left-hand tail of the distribution; accept the alternative that the dispersion is regular.

A visual means of testing the dispersion of a population is to construct a diagram like figure 1.4, using the following steps:

1. Calculate the mean number of individuals per unit, $\overline{N}$.
2. Using $\overline{N}$ for μ in the Poisson distribution (equation 1.1), calculate

the expected frequency or number of units containing 0, 1, 2, . . . individuals.

3. Plot on the same graph the expected frequencies from the results of step 2 and the observed frequencies of individuals per sample.

The two sets of figures may be compared statistically by a chi-square test of goodness of fit, a procedure described in most statistical texts. Examples of this method applied to plant populations are presented in Grieg-Smith (1964, p. 61) and Kershaw (1964, chap. 7).

Ex. 1.5. A test for random dispersion of a population. In 100 samples taken from square-meter plots the numbers of individuals per sample were counted. The data are summarized in the following table and are plotted as vertical bars in Figure 1.4B.

Number of Indivs.	Number of Samples	Number of Indivs.	Number of Samples	Number of Indivs.	Number of Samples
0	6	7	5	13	8
1	5	8	6	14	7
2	3	9	7	15	6
3	4	10	8	16	4
4	4	11	7	17	2
5	5	12	6	18	3
6	4			Total	100

$\overline{N} = 9.07$; $S = 5.01$; Coefficient of dispersion $= \frac{9.07}{5.01} = 1.81$

From equation (1.6), $\chi^2 = \frac{100 \times 25}{9} = 278$; d.f. $= 100 - 1 = 99$

With 99 degrees of freedom the probability of exceeding a χ^2 of approximately 135 is 1 percent. Since 278 is much greater than 135, the null hypothesis of random dispersion should be rejected at the 1 percent level (or higher), and the population judged to have an aggregated dispersion.

Another approach to testing for dispersion is the use of "Taylor's power law" (Taylor 1961). He found that for several species the sample variance, S^2, was a power function of the sample mean; that is, $S^2 = a\overline{N}^b$, or, $\log S^2 = \log a + b \log \overline{N}$. If the sample comes from a Poisson distribution or random dispersion, $\log a = 0$ and $b = 1$. If the dispersion is regular, b is less than 1; if aggregated, b exceeds 1. In the latter cases the value of a depends upon b and the sampling procedures. The steps in using Taylor's power law are as follow:

1. Obtain several samples, the samples differing in plot size, sampling technique, or habitat so that they will differ in mean number.

2. Calculate $\overline{N}$ and S^2 for each set.
3. Use linear regression methods for $\log S^2 = \log a + b \log \overline{N}$ to estimate b and to test whether or not b differs significantly from 1.

An example of this procedure is given by Elliott (1971). Taylor (1961) also describes how this power function may be used to find the best transformation to stabilize the variance before using statistical procedures such as analysis of variance. If $b = 1$, the square root transformation is best as stated in an earlier paragraph of this chapter.

Most natural populations whose dispersions have been studied were found to be aggregated. This alone does not tell us very much. An aggregated population may have its clumps dispersed randomly, regularly, or aggregated into clusters of clumps. Within each clump the individuals may be dispersed in any one of the three modes. Several methods have been tried to fit a frequency distribution to the data from an aggregated population; the negative binomial distribution has apparently been used most successfully. Mathematical approaches to the problem may be found in Pielou (1969) and in several papers by different authors in Patil, Pielou, and Waters (1971). Brief expositions of some methods are presented by Poole (1974, chap. 5). Grieg-Smith (1964) describes sampling procedures and graphical analysis which can reveal more about the patterns of aggregation. It is questionable, however, whether any one of these measures, used by itself, reveals enough ecological information to be worth the work of calculating it. Aggregation may result from patchiness of the environment, social behavior of the species, or reproduction with limited dispersal of the offspring, or from other causes. For significant conclusions, therefore, measurements of local variations in population density must be related to other characteristics of the population or to features of the environment.

2 Estimating Density by Specialized Methods

The mobility of animals, and especially their propensity to move away and hide from an observer, may prevent the estimation of their density by typical sampling methods. So some different procedures have been invented. One, the strip count, estimates the area in which the animals are counted by observations made during the count. Another, the removal method, uses a statistical procedure to estimate the total number in an area from a series of incomplete captures. Attempts have been made to estimate density by the distance between individuals. A different group of methods yields only an index of density or abundance, but this kind of estimate is adequate for many purposes. None of the methods described in this chapter are useful for estimating population dispersion.

2.1 THE STRIP COUNT

Strip counts are frequently used in studies of game populations because of the ease of their operation. An observer moves along a predetermined line and counts all the individuals observed on both sides. The variations in this method are in the ways in which the width of the strip is determined. In the simplest, only the animals within a certain distance are counted; this is no different from counting within an elongated sample plot, and it has the limitations of not knowing how many individuals were missed and of not being able to use the data from observations of animals beyond the limits of the plot. In the more sophisticated procedures, the size of the plot is determined by the observations. Several estimators have been proposed for this determination (Seber 1973, pp. 28–41); the method described below was originally proposed by Hayne (1949) and appears to be the most satisfactory.

The procedure, assumptions, and calculations can be described easily by a hypothetical example of a strip count for grouse. An observer walks a line of *L* meters in length. For each grouse observed he estimates or measures the distance, *D* meters, between himself and the bird when first seen. Each individual bird is assumed to have its own characteristic conspicuousness, flushing distance, or whatever leads to its discovery, and this is assumed to be independent of the direction of the observer's movement. A particular grouse, for example, is assumed to flush at a certain distance whether the observer is approaching directly or at an angle, and so if that bird is anywhere within *D* meters of the center line it will be counted. The preceding statement is the basic assumption of the method, and the calculations can be explained using figure 2.1. It is assumed that eight grouse have been observed, with discovery distances varying from D_1 to D_6; two birds were seen at D_2 and two at D_4, one bird at each of the other distances. In the figure the parallel lines at each distance delimit the

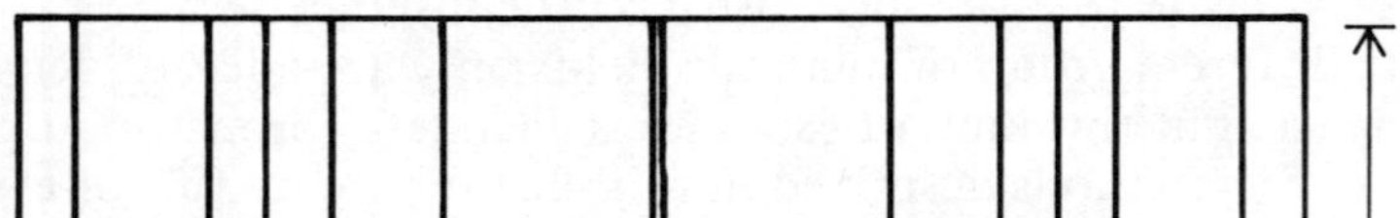

Figure 2.1. Strip count results. Each D is a discovery distance measured from the center line. See text for method of calculating population density.

whole area in which each individual bird could have been counted. The following information is contained in the figure:

There are:

1 "type 1" bird per ($L2D_1$) square meters = N_1/hectare;
2 "type 2" birds per ($L2D_2$) square meters = N_2/hectare;
1 "type 3" bird per ($L2D_3$) square meters = N_3/hectare;
2 "type 4" birds per ($L2D_4$)square meters = N_4/hectare;
1 "type 5" bird per ($L2D_5$) square meters = N_5/hectare;
1 "type 6" bird per ($L2D_6$) square meters = N_6/hectare.

In summary,

$$\frac{1}{2L} \Sigma_{i=1}^{6} \left(\frac{\text{number of type } i \text{ birds counted}}{D_i}\right) = \Sigma_{i=1}^{6} N_i/\text{hectare}. \qquad (2.1)$$

(Note: the conversion factor—10,000 sq. m. per hectare— is omitted.)

Ex. 2.1. Calculating density from a strip count of variable width. The length of the strip was 500 meters. The numbers of birds and the distances at which they were observed were: one bird at 16 meters, two at 24, one at 28, two at 32, one at 41, and one at 45 meters. By equation (2.1),

$$\text{estimated density} = \frac{1}{2 \times 500}\left(\frac{1}{16} + \frac{2}{24} + \frac{1}{28} + \frac{2}{32} + \frac{1}{41} + \frac{1}{45}\right)\left(\frac{10{,}000 \text{ sq.m.}}{\text{hectare}}\right)$$
$$= 2.9/\text{hectare}$$

This estimation assumes, as stated previously, that the discovery distance is independent of the observer's angle of approach. Eberhardt (1973) studied the effects of variations from this assumption and concluded that under a variety of conditions, equation (2.1) is an unbiased estimator and is therefore a robust estimator. An assumption made in all strip count estimations is that each discovery is independent of the others (e.g., that one bird's flushing does not cause another to do so). An analysis of data from strip counts of grouse (Sen, Tourigny, and Smith 1974) shows that this is not true, but no method has been proposed to allow for this lack of independence.

For a valid statistical estimate of the population density by this method, the lines or transects should be chosen at random and several replicates made to allow for the calculation of a mean and a standard deviation. An analysis of variance applicable to this kind of count is described by Schultz and Muncy (1957). Accuracy can be improved by stratifying the area by habitat and selecting random transects within each type.

Overton (1971, pp. 420–23) describes variations of the strip count as used for game birds. Emlen (1971) used a strip count to obtain a "coefficient of detectability" for each species of song bird and in each habitat, which could then be used to convert counts made along a transect into estimates of density.

Yapp (1956), in a more rigorous analysis of this method, has shown that the average speeds of the observer and of the animals should also be included in the calculations. If the animals were not moving, or are moving very slowly in comparison to the observer, these factors can be neglected. But if the animals are moving relatively rapidly, and the calculations described in a preceding paragraph—calculations which neglect velocity—are used, an overestimate of density will be obtained. Yapp's refinement has not been very useful because of the difficulty of measuring the velocity of animals in the field.

2.2 THE REMOVAL METHOD

When individuals are caught and removed from a population, and the catching operation is repeated again and again, the number of individuals caught each time is expected to decrease until the last one is removed. Usually only part of the population is removed, and the original number is estimated from the progressive decrease in catch size by the calculations described below. This method is adapted to estimating population size in a small "island" habitat (e.g., fish in a pond or insects in an isolated stand of plants) where the situation minimizes the escape of individuals from the disturbance of trapping or catching and the entrance of immigrants during the operation. The removal method has the obvious disadvantage that the population will be profoundly affected by the procedure. A modification that avoids this, based on marking rather than removing animals, will be described later.

The calculations of population size by the removal method depend upon these assumptions:

1. The probability of any particular individual being caught is the same for all, and is not affected by the operations; this includes the implication that there is equal effort of catching in each operation (but see second paragraph below for a modification that allows unequal effort).
2. There is no significant immigration, emigration, reproduction, or mortality exclusive of that caused by catching and removal.

If these assumptions are correct, the number caught in the ith operation, C_i, equals the probability of each individual being caught, c, times the number in the population at the time of that operation, N_i:

$$C_i = cN_i = c(N_1 - X_i) = cN_1 - cX_i, \qquad (2.2)$$

where N_1 is the number in the population at the time of the first catch and X_i is the number removed by all preceding catches (i.e., the cumulative catch up to but not including the *i*th catch). Equation (2.2) says that C_i is a linear function of X_i as shown in Figure 2.2, with the slope of the line being equal to $-c$ and the intercept on the vertical axis to cN_1, which is the constant of the regression. The line will intercept the horizontal axis at $X = N_1$, when zero catch would result because the whole population has been removed. This can be seen easily by reversing the scale on the horizontal axis, numbering it (in the example of figure 2.2) from 65 to 0, and labeling this axis "Number of animals remaining."

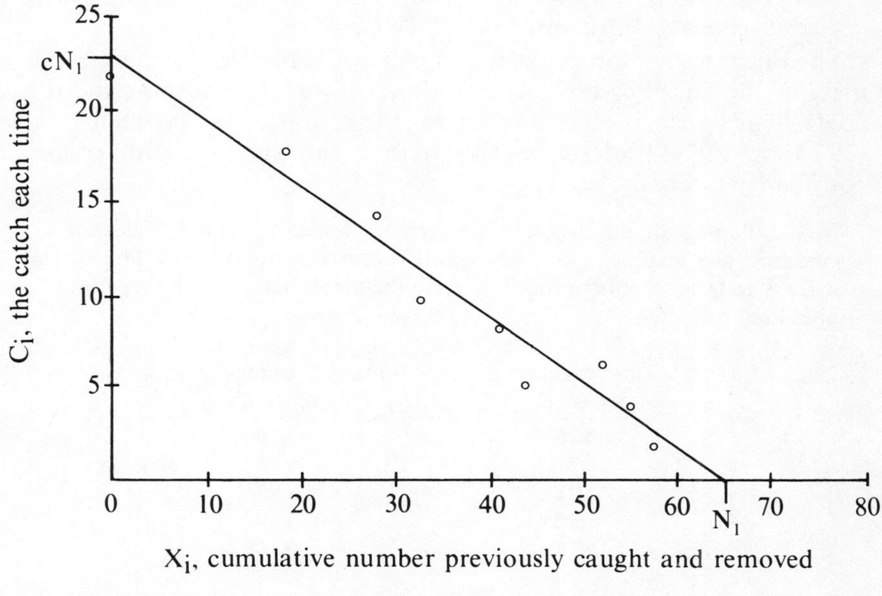

Figure 2.2. The removal method of estimating population size. The number caught each time, C_i, is a linear function of the number previously caught and removed.

Allowances can be made for variations in the effort of catching if these can be measured. The C_i of equation (2.2) is replaced with C_i/f_i where f_i is a measure of, for example, the "fishing effort." The catch per unit effort is then assumed to be a linear function of the number of animals remaining in the population.

Because of random errors and other variation, successive values of C_i or C_i/f_i probably will not fall exactly on a straight line. An objective way of estimating the position of the line is to use linear regression. The steps are:

1. Tabulate the number removed, C_i, or number removed per unit effort, C_i/f_i, at each catch.
2. Calculate and tabulate for each catch the cumulative number removed up to but not including that catch; $X_i = \Sigma_{j=1}^{i-1} C_j$.
3. Using X_i as the independent variable and C_i as the dependent variable, use linear regression to estimate the slope, $-c$, and the constant, cN_1, of the regression (equation 2.2).
4. Calculate the original number present from $N_1 = cN_1/c$ = (constant of regression)/(regression coefficient).

   N_1 can also be estimated by projecting the regression line until it reaches the horizontal axis, $C_i = 0$ and $N_1 = X_i$.

The linear regression calculations will not lead directly to an estimate of the sampling error of N_1 (they will for the constant and the coefficient by themselves but not for the ratio of the two). Seber (1973, pp. 297–99) describes this method and the estimation of a confidence interval.

Ex. 2.2. Population numbers estimated by the removal method. Fish traps were used in a small lake to catch and remove bass for eight days. The number of traps and the trapping procedure were the same each day. The results are tabulated.

Day i	Number removed each day, C_i	Cumulative number removed in preceding catches, X_i
1	125	0
2	72	125
3	91	197
4	82	288
5	55	370
6	90	425
7	48	515
8	40	563

Using the symbols of equation (2.2), the linear regression equation fitted to this data is

$$C_i = 111 - .115X_i,$$

from which $\hat{N}_1 = \frac{111}{.115} = 965$.

A useful variation of the removal method is that of marking and releasing the animals caught instead of permanently removing them. This is still treated as the same method because it is the "removal of unmarked" animals from the population. The procedure is well-adapted to the study of small mammal populations for the following reasons. Since trapping usually needs to be done over several days, if animals are actually removed from the area, empty habitat is created thus encouraging immigration from outside. The release of marked individuals in their home ranges, however, avoids the effect of artificial removal. Another advantage of this method in the study of small mammals is that the sizes of the home ranges can be estimated from retrappings. These dimensions may then be used to correct for the fact that the margin of the trapping grid will probably cut across individual home ranges, raising the question of exactly how much area is being sampled by the traps. There is no widely accepted method of making an adjustment for this cause of error; papers by French *et al.* (1971) and Smith *et al.* (1975) describe some ways in which this may be done.

The steps used in the "unmarked removal" method are the same as those given above, except that C_i is the number of unmarked animals caught in the *i*th catch (data on the number of marked animals caught in each catch are not used in this method). The cumulative catch, X_i, is as before the sum of the C_i, of all unmarked individuals, caught up to but not including the *i*th catch. Data on both the unmarked and marked individuals can be used to estimate population size by the repeated mark-recapture method described in chapter 3, section 3.2, where both methods are compared.

The regression estimate described above is not, from the viewpoint of theoretical statistics, the best means of estimating population number from removal data. A thorough discussion of the subject is by Zippin (1956, 1958); he presents an alternative system of calculation, using graphs to reduce some of the computation. Southwood (1966) and Seber (1973) summarize Zippin's procedures and reproduce the graphs. Practically, however, there is usually so much random variation in data acquired by the removal method that little accuracy is gained by the use of calculations other than those of regression, which are the most flexible and the easiest to use. A different approach to estimation of population number by regression is provided by Skellam (1962), who suggests transforming the time scale and using logarithms of the cumulative catch; this method, apparently little used in field studies, would allow an easier estimate of a confidence interval.

Zippin (1956) states that the precision of the estimate by the removal method depends upon the proportion of the population that is eventually removed; good precision requires that this proportion be large. It was his opinion that the mark-recapture method of estimating population size (described in the next chapter) will usually give better results for the same amount of effort. This, however, will depend upon whether the field conditions and the behavior of the animals better satisfy the requirements of one method or the other.

Applications of the removal method to fish populations are described in Ricker (1975, chap. 6); to insects in Southwood (1966, chap. 7); and to various groups by Seber (1973, chap. 7). Each of these authors also describes some modifications in the calculations.

2.3 THE NEAREST-NEIGHBOR METHOD

In denser populations individuals are obviously closer together, and this fact suggested estimating density from measurements of the distances between individuals. The usual procedure is to select one individual at random and then to measure the distance to its nearest neighbor. This "nearest-neighbor distance," measured a number of times, can be used to estimate density if the population is randomly dispersed, or to determine the dispersion if the density is known. Variations of this method have been most used in plant ecology (Grieg-Smith 1964, Mueller-Dumbois and Ellenberg 1974). Some applications to insects are described by Southwood (1966, pp. 39–43), and Blackith (1958) describes an interesting test of the method, also on insect populations. It has been little used in animal ecology because of the difficulty of measuring nearest-neighbor distance in mobile organisms and because other methods provide better statistical information.

2.4 RELATIVE DENSITY AND INDEXES OF DENSITY

Relative density, or relative abundance, is a measure based on counting individuals with respect to units other than space. Examples are catch per unit effort, catch per trap-night, individuals observed per hour, and sweep-net samples. An *index of population density* is based on some sign of animals, such as tracks, feeding sign, and droppings or feces. To be useful these indexes must be some increasing, and preferably a linear, function of the true density of the population.

Measurements of this kind may be much easier and more economical to make than measurements of absolute density, and they may be perfectly satisfactory for the solution of some research problems. If the question that needs to be answered is whether or not there is a significant difference between two areas—or at different times, such as before and after an experiment—relative measurements may give adequate data for the tests with less effort.

Measurements of relative density or indexes are, on the other hand, of little value if the densities of two different species need to be compared. Each group has its own behavior and conspicuousness, and it is very unlikely that relative and absolute densities will be related in the same way for even closely related species. The same applies to estimates of one species at different seasons or of different ages.

Southwood (1966, chaps. 7 and 8) describes many different relative measures applicable to insects. Ricker (1975, p. 21 and chap. 6) discusses and gives examples of the use of catch per unit effort in commercial fisheries, and the validity of this general method is further discussed by Cushing (1968, pp. 68–76) and by Watt (1968, pp. 213–24.). Beverton and Holt (1957) published a method which depends upon catch per unit effort but which can yield estimates of absolute population size and other characteristics. The data required are catch per unit effort of each age class of fish over a sequence of years. The calculations are explained by Cushing (1968, pp. 61–63) and by Watt (1968, pp. 211–13). Regier and Robson (1967, pp. 53–56) discuss the method and conclude that the necessary assumptions for making the calculations are rarely met and thus they question the usefulness of the procedure.

The subject of relative measurements of bird populations has been reviewed by Kendeigh (1944). He also describes the method often referred to as a breeding bird census, but which is really a kind of index since it counts territorial males. Observations of singing males, intraspecific fights or displays, and other activities are mapped, and then the locations are interpreted to estimate the boundaries of individual territories. The procedure has been standardized (Hall 1964, Svennson 1970). Best (1975) discusses sources of error in the method, especially the variations resulting from different interpretations of the maps. A critical review of the breeding bird census has been written by Berthold (1976) who points out that errors frequently have been large and that there is no easy way to determine the precision of the estimate. A standardized roadside count of breeding birds that is widely used in the United States is described by Aldrich and Robbins (1970) and Robbins and van Velzen (1970). A modification of this method that gives good relative counts of birds in other habitats consists of walking a prearranged route, stopping for a three-minute period to record all individual birds heard or seen from that point, then walking for three minutes to the next station where another three-minute count is made. The three minute walk is usually long enough to prevent counting the same individual bird at two stations. Records are kept separately for each station. Several methods used for relative counts of game birds and mammals are described by Overton (1971, pp. 414–16, 424–32). Some of these methods resemble strip counts described in the preceding section except that no attempt is made to measure the area within which the animals are observed. The relative abundance of amphibians and reptiles can be measured by timed searches. A representative habitat is chosen, and the observer makes several searches within the habitat, each for a predetermined period (e.g., fifteen minutes), recording the species and individuals found in that time. Relative estimates of small mammal densities are included in the methods described by Smith *et al.* (1975).

An analysis of variance applicable to relative counts of animals is described by Schultz and Muncy (1957). In our earlier discussion on statistical tests applied to sampling data (chap. 1), however, the advantages of using the median test are described, and this nonparametric test appears to be a safer procedure than analysis of variance with its restrictions.

2.5 PERSPECTIVE

Although additional methods of estimating population size are discussed in the next two chapters, this is an advantageous point to consider the subject from two viewpoints; the first is that of the purpose for which an estimate of density or number is needed, and the second is that of the characteristics of the animals being studied.

Estimate of population size may be needed for management purposes, such as determining how many can beneficially be harvested or removed from the population, or the potential danger from an insect pest. The total number in a population can be estimated by any of the methods of sampling, including strip counts, and by the removal method (although this may radically reduce the population). Population number can also be estimated by the mark-recapture method described in chapter 3 and by change in ratio methods included in chapter 4.

Another purpose of making estimates is to compare either numbers or density between two populations, in different habitats or at different times. Any of the methods mentioned in the preceding paragraph may be used to acquire data for statistical tests providing that replicates can be obtained for estimating the variance or for use in a nonparametric test. Indexes of abundance or measurements of relative abundance may be used for the same purpose, and measures of this nature are often much easier and cheaper to obtain.

Animals which are sessile or very slow moving may be counted in sample plots or similar units. Examples are many species of aquatic organisms which are anchored to the bottom, soil animals, and insect larvae. These are well adapted to being studied by standard sampling methods which can yield much information about the population.

Mobile animals usually cannot be counted in well-defined sampling units, so recourse can be made to a strip count to estimate their density. Statistical methods for analyzing data from these counts are not as well developed as are those applicable to standard sampling methods. Indexes of abundance are often of more value with these species because they are easier to obtain and therefore to replicate, and they often supply the information that is needed in research or for management decisions. If the animals can be captured and marked for future identification in such a way that their behavior after release is not modified, mark-recapture methods may be used to estimate numbers.

Some animals, such as fish and nocturnal mammals, cannot be easily counted in their natural habitats, but they may be caught, and under certain conditions this allows estimates of population size by the removal method. This is probably not economical unless the catching and removal is being done for some other reason, such as by a commercial fishery. The same generality is probably true for estimates made by the change in ratio method described in chapter 4.

About 160 citations to papers on methods described in this chapter are included in a bibliography on quantitative ecology (Schultz *et al.* 1976, pp. 265–76).

3 Mark-recapture Methods of Measuring Population Characteristics

In the simplest mark-recapture procedure, a number of animals are captured, marked so that they can be identified, and released into the population. Later, a sample is taken in which some of the individuals are marked and the others are not. The symbols used here and throughout this chapter are:

N: number in the population, to be estimated;
M: known number of marked animals in the population;
C: total number captured in the sample;
R: the number recaptured (i.e., the number of marked animals in the C number captured).

The proportion of marked animals in the sample will, if sampling is random, estimate the proportion of marked animals in the entire population; R/C estimates M/N (see figure 3.1). Since R, C, and M are known, N can be calculated by MC/R.

This estimate has been called the Petersen method, after a marine biologist who marked fish in a population study, or the Lincoln index, from Lincoln's use of data on the number of banded waterfowl to estimate their populations. However, the best name appears to be the "mark-recapture method." The procedure has been widely used, criticized, and analyzed; the literature on the subject is enormous. Cormack (1968) offers a good history and review of the literature with emphasis on the use of the method in fisheries research. Seber (1973) devotes four chapters to the mathematics of the method and its variations, and he gives numerous examples; this is to date the most thorough mathematical review of the subject. Tepper (1967) presents a bibliography listing nearly 300 titles; its second part includes about 100 titles on the method as applied to many different kinds of animals. Additional and more recent titles are included in a bibliography assembled by Schultz and others (1976, pp. 253–64).

The following sections will cover the single mark-recapture estimate, repeated mark-recapture using a single mark, multiple mark-recapture procedures, and methods of marking animals.

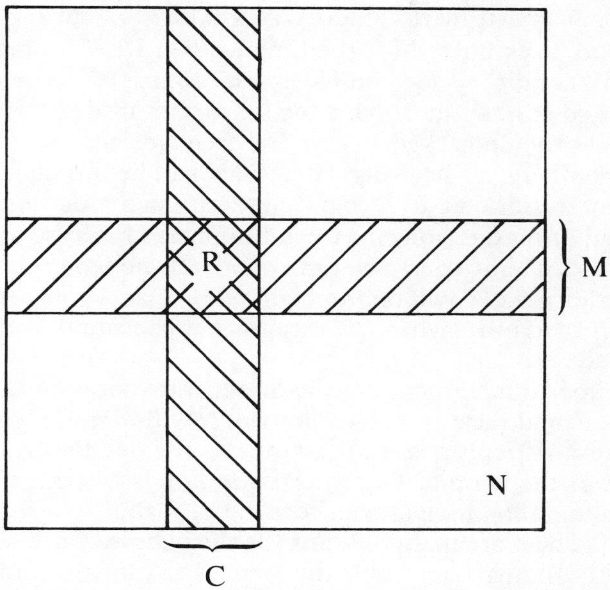

Figure 3.1. Single mark-recapture. The areas in the figure are proportional to the numbers of individuals. The total area represents the population size, N, of which M individuals have been previously marked. The random sample of C individuals includes R that were marked, and the proportion R/C estimates the proportion M/N.

3.1 SINGLE MARK-RECAPTURE

The single mark-recapture procedure has been outlined above. If MC/R is to be a reliable estimate of the population number, certain conditions must be met. Some of these depend upon whether N is considered to be the number in the population (a) when the marked animals were released, or (b) when the recapture was made. The first two conditions—(1) and (2)—apply to both cases.

1. The sample at the time of recapture must be random with respect to marking; that is, the fact that an individual has been marked and released does not affect the probability of its recapture. This does not mean that both the marking and the recapture must be random with respect to the entire population;

it does mean that the two operations must be random with respect to each other. Only if this is true is the ratio R/C an unbiased estimate of M/N.

2. The marks must endure and marked animals must always be recognized. (If marks are lost, $\hat{N}$ will be too high.)

If N is to be estimated for the time when marked animals were released (a), conditions (3) and (4) are necessary.

3. Marked animals must have the same rates of death and emigration as the unmarked so that M/N remains constant. (If the marked animals have higher rates, $\hat{N}$ will be too high.)
4. There must be no unrecognized recruitment—by immigration, birth, or maturation—between the times of release and recapture; recruits will of course be unmarked. (If unrecognized recruitment occurs, $\hat{N}$ will be too high.)

If N is the number when the recapture was made (b), condition (5) is required.

5. Marked animals must not die or emigrate between the times of release and recapture. (If some do, $\hat{N}$ will be too high.)

The most difficult condition to meet is the one first listed, the randomness of the sample. The crucial question is whether the capturing and marking of an individual affect the probability of its being recaptured. There are many reasons for the behavior or distribution of the marked animals not being the same as that of the unmarked, and therefore for the catchability of the two groups not being equal. The data from the single mark-recapture process provide no way to test for equal catchability, so the randomness of the sample can be evaluated only by knowledge of the animals and their habits. Randomness can be increased by using one method for catching animals to be marked and a very different method for the recapture; for example, quail can be caught by baited traps and then sampled by shooting in randomized locations. Other methods of "recapture" that are obviously different from the first capture are poisoning, examining victims of natural mortality for the presence of marks, and identifying visibly marked individuals without catching them.

In some species—fish are good examples—groups differing in size, age, or some other characteristic differ in their catchability. In these cases the recapture sample will be random only within each group. Consequently, mark-recapture estimates should be made only within the group; or in other words, data from different groups should not be lumped. Another way of estimating population size when subpopulations differ in catchability is described by Overton (1971) in what he calls a "stratified Lincoln Index."

Considering the other conditions listed above that are necessary for a good mark-recapture estimate, beginning with number 2, we note that if any one of these conditions is not met, the population estimate will be too high, so in practice $\hat{N}$ tends to be positively biased. Another factor adding to this positive bias is that there is always a probability that the recapture sample will contain no marked individuals, in which case MC/R equals infinity; data from such samples should not be used.

The validity of some of the assumptions can be tested. Tests for the loss of marks are described in section 3.5 of this chapter. Emigration of marked animals from the area may be detected by trapping outside the boundaries of the population being studied. Recruitment of unmarked individuals to a fish population may be recognized by methods described by Ricker (1975, pp. 82–86) and by Robson and Flick (1965). In a different procedure, the multiple mark-recapture which is to be described, gains or losses in the population can be estimated. Further comments, therefore, will be deferred until the methods can be compared near the end of this chapter.

The accuracy of the population estimate by the single mark-recapture method, assuming that all conditions are met, is determined by probability or chance. The effects of chance can be calculated by assuming that the sampling is to be performed many times under identical conditions, with N, M, and C constant. There will be a distribution of $\hat{N}$ determined by probability, and the expected results can be calculated to give an expected or mean value of $\hat{N}$ and an expected value of the variance of $\hat{N}$. The expected results will depend upon which is assumed to be the proper theoretical distribution (e.g., binomial, hypergeometric, etc.; see Cormack [1968] for a discussion of this). Bailey (1951, 1952) assumes the binomial distribution and gives the formulas that are presented below. These are in two forms because of the positive bias of $\hat{N}$; the "small sample" form is less biased and only slightly more difficult to compute.

	Expected $\hat{N}$	Expected Variance	
Large Sample	$\frac{MC}{R}$	$\frac{M^2 C(C - R)}{R^3}$	(3.1)
Small Sample	$\frac{M(C + 1)}{(R + 1)}$	$\frac{M^2 (C + 1)(C - R)}{(R + 1)^2 (R + 2)}$	(3.2)

To complete the argument started above, if the sampling could be performed many times under identical conditions, the many calculated values of $\hat{N}$ would be expected to have (with a slight bias) the mean and variance predicted by these formulas.

Ex. 3.1. Single mark-recapture estimate. Two hundred twenty insects were caught, marked, and released. A few hours later 240 were caught, of which 16 individuals were marked.

Using large sample equations (3.1):

$$\hat{N} = \frac{220 \times 240}{16} = 3300$$

$$\hat{\sigma}^2 = \frac{220^2 \times 240 \times (240 - 16)}{16^3} = 635{,}250$$

$$\hat{\sigma} = 797.$$

Using small sample equations (3.2):

$$\hat{N} = \frac{220 \times 241}{17} = 3119$$

$$\hat{\sigma}^2 = \frac{220^2 \times 241 \times (240 - 16)}{17^2 \times 18} = 502{,}273$$

$$\hat{\sigma} = 709.$$

A confidence interval for $\hat{N}$ may be calculated by using equations (3.1) or (3.2) with tables or charts based on the binomial distribution such as those appearing in Overton (1971). A more convenient way of calculating this confidence interval is to use the reciprocal $1/N$ which is normally distributed (Cormack 1968); $1/N$ is estimated by R/MC, and its variance by

$$\text{Variance of } \widehat{\frac{1}{N}} = \frac{R(C - R)}{M^2 C^3}. \qquad (3.3)$$

The steps for calculating a 95 percent confidence interval by this approach are:

1. Calculate $\widehat{1/N}$ and its variance from equation (3.3).
2. Calculate $\hat{\sigma}$, the square root of the variance and multiply by 2.
3. Subtract $2\hat{\sigma}$ from $\widehat{1/N}$ to obtain the lower limit, and add $2\hat{\sigma}$ to $\widehat{1/N}$ to obtain the upper limit, of the 95 percent confidence interval for $\widehat{1/N}$.
4. Invert $\widehat{1/N}$ to obtain $\hat{N}$, and invert the two limits to obtain the upper and lower limits for $\hat{N}$.

If an interval is desired for other than 95 percent confidence, the multiplier to use in step (2) can be obtained from a table of the normal distribution.

Ex. 3.2. Calculating a 95 percent confidence interval for a single mark-recapture estimate. Using the data of the previous example, a confidence interval can be calculated from equation (3.3).

$$\widehat{\frac{1}{N}} = \frac{16}{220 \times 240} = .000303$$

$$\hat{\sigma}^2 = \frac{16\,(240 - 16)}{220^2 \times 240^3} = 5.357 \times 10^{-9}$$

$$\hat{\sigma} = .000073$$

$$\frac{1}{N} + 2\hat{\sigma} = .000303 + .000146 = .000449; \frac{1}{.000449} = 2227$$

$$\frac{1}{N} - 2\hat{\sigma} = .000303 - .000146 = .000157; \frac{1}{.000157} = 6369$$

$$\hat{N} = \frac{1}{.000303} = 3300,$$ with a 95 percent confidence interval from 2227 to 6369.

The preceding discussion of the accuracy of a single mark-recapture estimate has been based on the hypothesis that C, the number in the recapture sample, was constant. Fixing the size of this sample is called the "direct" method by Bailey (1951, 1952). Another option, called the "inverse" method, is to fix R (i.e., to sample until a predetermined number of marked individuals are caught and then to stop.) Then the formulas for $\hat{N}$ and its variance are:

$$\hat{N} = \frac{C\,(M + 1)}{R} - 1; \text{ Variance of } \hat{N} = \frac{(M - R + 1)\,(\hat{N} + 1)\,(\hat{N} - M)}{R\,(M + 2)}. \quad (3.4)$$

The inverse method has the statistical advantages that $\hat{N}$ is unbiased and that the precision of the estimates can be partially controlled by choosing the size of R. However, it is not always possible to do this in practice, and in fact the method has seldom been used. In most recaptures in the field the sample has been as large as could economically be caught; this is not exactly the direct method but is near to it, and the formulas for that method should be used.

3.2 REPEATED MARK-RECAPTURE

The procedure in the repeated mark-recapture method is to mark and release a number of animals; to capture a sample, counting the number of marked and unmarked individuals in the sample, marking all of the previously unmarked, and releasing all of the sample to the population; and then to repeat this sampling and marking several times. The same mark may be used for all individuals. As more samples are taken, the proportion of marked animals in the population will increase as will the expected proportion of recaptures in each sample (see figure 3.2).

The conditions necessary for a reliable estimate of the population size are conditions 1 and 2 stated in the preceding section, and furthermore it is assumed that the population is closed throughout the period of sampling, with no mortality or recruitment. This is a more stringent condition than for the single mark-recapture estimate, but the repeated sampling produces larger numbers of marked and recaptured individuals which is an advantage.

The following symbols will be used:

M_i: the number of marked individuals in the population at sample i, equaling the sum of all individuals previously marked and released.

C_i: the total number of individuals caught in sample i.

R_i: the number of marked individuals in sample i.

Under the assumptions of this method the expected value of $N = M_1C_1/R_1 = M_2C_2/R_2 = M_iC_i/R_i$. Schnabel (1938) showed that a simple approximation to the maximum likelihood estimate is:

$$\hat{N} = \frac{\Sigma\,(M_iC_i)}{\Sigma R_i}. \tag{3.5}$$

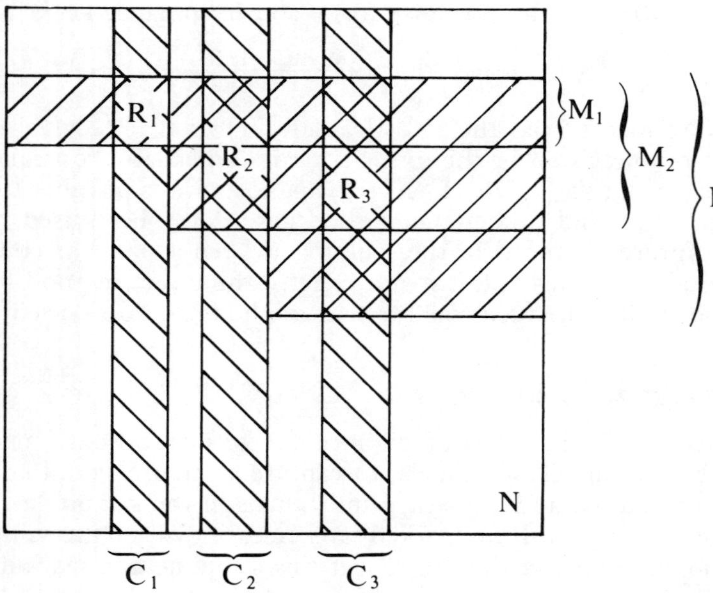

Figure 3.2. Repeated mark-recapture. The relations are the same as in Fig. 3.1. M_1 is the number originally marked and released and C_1 is the subsequent capture. All the unmarked individuals in C_1 are marked and released to increase the total of marked animals to M_2. The same steps are repeated at each sample. Each R_i/C_i estimates M_i/N. A weighted average of $\hat{N}$ is provided by equation (3.5).

Because of her contribution, this procedure is often called the "Schnabel method."

For calculating the confidence intervals for this estimate, Ricker (1975, pp. 96–97) recommends estimating the reciprocal of $\hat{N}$, by inverting equation (3.5), and estimating the variance:

$$\text{Variance of } \widehat{\frac{1}{N}} = \frac{\Sigma R_i}{(\Sigma M_i C_i)^2}. \quad (3.6)$$

The values of $1/\hat{N}$ are approximately normally distributed so that confidence limits may be calculated for $\hat{N}$ by the steps described in the preceding section. The normal approximation may be avoided by using the method and table of Chapman and Overton (1966), who also describe a method of testing for significant differences between two estimates.

Ex. 3.3. Repeated mark-recapture estimate of population size and calculation of a confidence interval. Six successive seining operations caught fish which were marked and released. The data and the steps in the estimation of population size are tabulated.

i	M_i	C_i	R_i	Newly marked & released	M_iC_i
1	0	44	0	44	0
2	44	53	4	49	2332
3	93	58	6	52	5394
4	145	47	11	36	6815
5	181	52	17	35	9412
6	216	46	18	28	9936
TOTALS			56		33889

$$\hat{N} = \frac{33889}{56} = 605; \quad \widehat{\frac{1}{N}} = .00165 \quad \text{(eq. 3.5)}$$

$$\hat{\sigma}_{1/N} = \sqrt{\frac{56}{33889^2}} = .00022 \quad \text{(eq. 3.6)}$$

$$\widehat{\frac{1}{N}} + 2\hat{\sigma} = .00165 + .00044 = .00209; \quad \frac{1}{.00209} = 478$$

$$\widehat{\frac{1}{N}} - 2\hat{\sigma} = .00165 - .00044 = .00121; \quad \frac{1}{.00121} = 826$$

$\hat{N}$ = 605 with a 95% confidence interval of 478 to 826.

Another method of estimating N from repeated mark-recapture data is that of Schumacher (Schumacher and Eschmeyer 1943). In these calculations the ratio R_i/C_i is considered to be a linear function of M_i. That is, $R_i/C_i = (1/N)M_i$, and $1/N$ is a coefficient of a linear regression. This coefficient and its standard deviation can be estimated by linear regression methods applicable to a regression line that passes through the origin. The Schnabel and Schumacher methods give very similar estimates from the same data; the calculations of the Schnabel method are simpler.

The repeated mark-recapture method uses the same operations and provides the same data as does the "removal of unmarked" method described in chapter 2, section 2.2. Both methods make the assumption that the population is closed during the procedure. The assumptions made for the estimates, however, are slightly different. The repeated mark-recapture method assumes that all individuals, whether marked or unmarked, are equally likely to be caught. The other procedure assumes only that all unmarked animals at the time of any catch are equally likely to be caught; they do not need to have the same probability of capture as those marked. This can be seen by using equation (2.2) to show that the slope of the line, c, is the ratio (number of unmarked caught)/(number of unmarked remaining in the population) at any time. The two methods, therefore, can be used to make two somewhat independent estimates of population size. If there is a discrepancy between the estimates, the student should examine the data for differences in catchability and for possible chance errors arising from small numbers in the data. Example 3.4 uses the same data as 3.3 for an estimate by the "removal of unmarked" method.

Cormack (1968) and Seber (1973) review the literature on the repeated mark-recapture method, and both discuss possible modifications of the estimates and means of testing some of the basic assumptions. Overton (1971) describes one modification that may frequently be necessary when known numbers of individuals are removed from the population such as by being killed in the sampling and marking process.

The choice between using a single or a repeated mark-recapture procedure is not always clear. The first has fewer assumptions and is less affected by conditions deviating from the assumptions; if all the

Ex. 3.4. Using repeated mark-recapture data in a "removal of unmarked" estimate. The data of example 3.3 will be used; the column headed M_i is the cumulative number "removed" up to time i, the X_i of equation (2.2), and the column headed "Newly marked and released" is the number of unmarked caught each time, the C_i of equation (2.2). Equation (2.2) and the linear regression fitted to the data are:

$$C_i = cN_0 - cX_i = 50.66 - 0.088X_i$$

$$N_0 = 50.66/0.88 = 576$$

conditions are met, or variations from them measured in the repeated sampling, the second should give better estimates. The choice should be based largely on the investigator's knowledge of the species and its behavior. If the opportunity exists for repeated sampling, and if a method can be used for marking each animal with an individual mark, then the multiple mark-recapture described in the following section is the better procedure, in that it does not assume a closed population.

3.3 MULTIPLE MARK-RECAPTURE

The mark-recapture estimate that depends upon the fewest assumptions and also yields the most information requires a series of three or more capturing operations and a method of marking that will reveal the history of capture of each individual. This may be done by marking each animal with a uniquely numbered tag, but a study of the data used in the calculations will show that other marking methods may be possible. The basic assumptions are that capture is random with respect to marked and unmarked individuals and that all previously marked individuals can be recognized. The population is not assumed to be closed; individuals may leave or enter the population between the times of capture, and the calculations yield estimates of the survival and recruitment rates.

Multiple mark-recapture data are best analyzed by calculations published almost simultaneously by Jolly (1965) and Seber (1965). The Jolly-Seber method is more general and powerful than any previously published and is preferred over others (earlier methods are summarized by Cormack 1968). The formulation by Jolly (1965) is slightly more general and is used here. His formulas take into account the loss of individuals at the time of marking (a frequent occurrence in practice), removal of animals by death or emigration, and recruitment by reproduction or immigration. The calculations do not depend upon the captures being made at regular intervals. The population model is stochastic, meaning that some of the quantities (such as the number of marked and unmarked animals alive at a particular time) are subject to random variation. Despite this complicating approach to the problem, the results after much tortuous algebra (see Jolly 1965) are surprisingly simple and intuitively understandable. The symbols used in the following explanation are changed from those of Jolly to be consistent with the rest of this book.

Organization of the Data

The symbols referring to the data are as follows:

C_i: the number, marked plus unmarked, caught in sample i, with i ranging from 1 (time of first marking) to z (time of last capture).

R_{hi}: number of marked animals in sample i that were last or most recently caught in sample h, with h ranging from 1 to $(z - 1)$, $h < i$. For example, if an individual were caught in samples 1, 2, and 5, it would be counted in the R_{12} and R_{25}, but not in the R_{15}.

F_i: number of marked animals released from sample i; F_i is equal to or less than C_i.

Table 3.1 sets forth how these data are organized.

Table 3.1. Organization of Data for Multiple Mark-Recapture Estimate from a Sequence of Seven Captures (see text for explanation).

		h						
i	C_i	*1*	*2*	*3*	*4*	*5*	*6*	F_i
1	C_1	-	-	-	-	-	-	F_1
2	C_2	R_{12}	-	-	-	-	-	F_2
3	C_3	R_{13}	R_{23}	-	-	-	-	F_3
4	C_4	R_{14}	R_{24}	R_{34}	-	-	-	F_4
5	C_5	R_{15}	R_{25}	R_{35}	R_{45} $\underline{R_{.5}}$	-	-	F_5
6	C_6	R_{16}	R_{26}	R_{36}	R_{46}	R_{56}	-	F_6
7	C_7	R_{17}	R_{27}	R_{37}	R_{47} $\underline{U_5}$	R_{57} $\underline{V_5}$	R_{67}	F_7

The following symbols refer to specific sums of the R_{hi}; components of each sum for $i = 5$ are enclosed in a separate box of dotted lines in table 3.1, and the symbol for each sum is the underlined figure within each box.

$R_{.i}$: the sum of all marked animals in sample i.

U_i : the number of marked animals *not* caught in sample i but caught subsequently, therefore known to be alive at time i.

$$U_i = \Sigma_{h=1}^{i-1} \Sigma_{j=i+1}^{z} R_{hj}.$$

V_i: the number of individuals released in F_i that are caught subsequently, and therefore known to be alive at time $i + 1$.

$$V_i = \Sigma_{j=i+1}^{z} R_{hj},$$

for $h = i$.

Additional symbols are:

N_i: number in population at time i.

M_i: total number of marked animals existing in population at time i (not necessarily the total previously marked).

a_i : proportion of marked animals in the population, M_i/N_i.

s_i : probability of survival of individuals from time i to $i + 1$.

B_i : number recruited (by either reproduction or immigration) to the population in the interval i to $i + 1$.

Jolly (1965) developed formulas for three situations: when both removal and recruitment occur, when removal but not recruitment occurs, and when recruitment but not removal occurs. Only the first two will be discussed here because the third situation is unlikely.

Estimates of Population Size, Survival, and Recruitment in the General Case

A quotation from Jolly (1965, p. 238) will serve to explain the major part of his solution (symbols have been changed to those listed above):

> The capture-recapture problem is therefore solved if we can estimate two things for each relevant point in time, namely the *proportion*, a_i, and the actual *number*, M_i, of marked animals in the population. The first of these is readily estimated by finding the proportion of marked animals in a random sample of the population. The *number* of marked animals can be thought of as consisting of the observed number captured, $R_{.i}$, plus the number not captured, $(M_i - R_{.i})$. Immediately after time i there are two groups of marked animals, the $(M_i - R_{.i})$ and the F_i animals just released. Of the former U_i are subsequently caught; of the latter V_i are subsequently caught. Therefore, since the chances of recapture are assumed to be the same for both groups,

the probability of recapture is $U_i/(M_i - R_{.i}) = V_i/F_i$, and solving this equality for M_i gives $M_i = (F_iU_i/V_i) + R_{.i}$. Figure 3.3 is a graphic explanation of these relations.

M_1 and a_1 will of course be zero, since there are no marked animals in the population at the first capture. Examination of the formula for M_i will show that M_z, at the time of the last sample, cannot be estimated because neither U_z nor V_z has any meaning. Therefore N_i cannot be estimated for the times of the first and last samples. This limitation is indicated in the following formulas which are Jolly's maximum likelihood estimators.

$$\hat{a}_i = \frac{R_{.i}}{C_i}, \qquad i = 2, \ldots z \tag{3.7}$$

$$\hat{M}_i = \frac{F_i U_i}{V_i} + R_{.i}, \qquad i = 2, \ldots z - 1 \tag{3.8}$$

$$\hat{N}_i = \frac{\hat{M}_i}{\hat{a}_i}, \qquad i = 2, \ldots z - 1 \tag{3.9}$$

$$\text{Variance of } \hat{N}_i = N_i (N_i - C_i) \left[\frac{M_i - R_{.i} + F_i}{M_i} \left(\frac{1}{V_i} - \frac{1}{F_i} \right) + \frac{1 - a_i}{R_{.i}} \right]. \tag{3.10}$$

The above formula is the variance of $\hat{N}_i$ for a given N_i, and does not consider the stochastic nature of N_i. Jolly (1965, pp. 237–38) gives the equation for the full variance including both the errors of estimation and the variance of the parameter N_i, but its calculation is troublesome. In using equation (3.10), $\hat{N}_i$ must be substituted for N_i, and $\hat{M}_i$ for M_i.

An important assumption of these calculations is that all animals are equally likely to be captured, which includes the idea that the

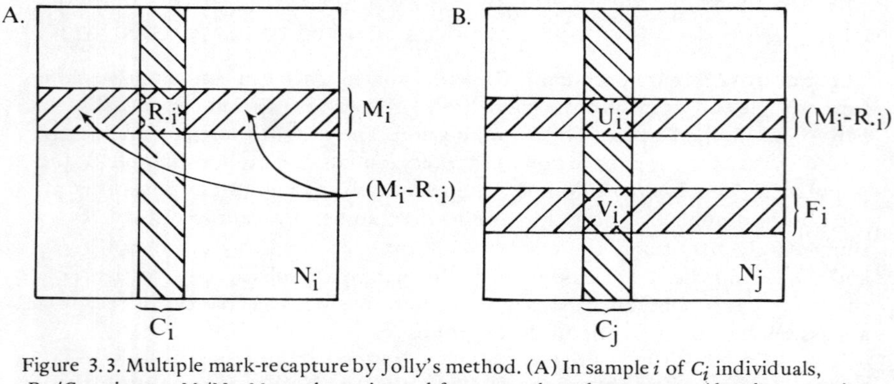

Figure 3.3. Multiple mark-recapture by Jolly's method. (A) In sample i of C_i individuals, $R_{.i}/C_i$ estimates M_i/N_i. M_i can be estimated from samples subsequent to i by the reasoning illustrated in the second figure. (B) U_i are the marked individuals not caught at time i but caught later (at time j), and V_i are the recaptured individuals from the F_i marked animals released at time i. These two groups should be sampled randomly at time j, so equal proportions should be included in the sample, or $U_i/ M_i - R_{.i} = V_i/F_i$. The only unknown in this equation is M_i.

probability of recapture is not affected by an individual's previous history of capture. It is possible, for example, that animals just released, F_i, would be less likely to be caught than animals not caught at time i. In which case V_i would be smaller than expected and therefore $\hat{M}_i$, estimated from equation (3.8), would be larger than the true

M_i. This would lead to an overestimate of N_i. An extreme overestimate can be avoided by recognizing that $\hat{M}_i$ cannot logically exceed the total of individuals marked and released. This is a point that the operator should check, and he should, if there is a contradiction, substitute the known total of marked individuals for an erroneous estimate.

The maximum likelihood estimate of the survival rate is

$$\hat{s}_i = \frac{\hat{M}_{i+1}}{\hat{M}_i - R_{.i} + F_i}, \qquad i = 1, \ldots z - 2. \tag{3.11}$$

The logic of this is that immediately after the F_i animals are released, the number of marked animals in the population is $(M_i - R_{.i})$ plus the F_i, and M_{i+1} of these are alive at time $i + 1$.

The estimate of the number recruited to the population between times i and $i + 1$ is

$$\hat{B}_i = \hat{N}_{i+1} - \hat{s}_i(\hat{N}_i - C_i + F_i), \qquad i = 2, \ldots z - 2. \tag{3.12}$$

Immediately after time i, the number of animals in the population is $(N_i - C_i + F_i)$, and s_i is the proportion of these that survive to time $i + 1$; the difference between the number surviving and the number in the population at time $i + 1$ is the number recruited by reproduction or immigration. Formulas for the variances of $\hat{s}_i$ and $\hat{B}_i$ are complicated; they can be found in Jolly's paper (1965, equations 27).

The calculation of confidence intervals from Jolly's estimators, $\hat{N}_i$ and the variance of $\hat{N}_i$, is not really precise. These are maximum likelihood estimators, and therefore it might be assumed that $\hat{N}_i$ has an approximately normal distribution for large samples. But $\hat{N}_i$ is biased (Gilbert 1973), usually negatively, and there is a positive correlation between $\hat{N}_i$ and its variance (Manly 1971, Roff 1973a). As a result, underestimates of N_i appear to be more accurate than they really are. A rule of thumb suggested by Roff (1973a) is that an estimate should be considered "reliable" only if the standard deviation is less than 5 percent of $\hat{N}_i$, and then to take the 95 percent confidence interval as $\hat{N}_i \pm 0.1\hat{N}_i$ (because if $S < .05N$, the interval $[\hat{N} \pm .1\hat{N}]$ will be greater than the interval $[\hat{N} \pm 2S]$, which is the usual 95 percent confidence interval). These difficulties can be reduced in a practical way by increasing the efficiency of catching animals, making the probability of recapture as large as possible, as this will reduce both the bias and the variance of the estimate.

Ex. 3.5. Estimates of population size, survival, and recruitment from multiple mark-recapture data. Hypothetical data from a multiple mark-recapture operation are tabulated below. $\hat{N}_i$ and its variance and standard deviation are calculated for $i = 4$ and $i = 5$, and the survival and recruitment are calculated for the interval between these two times.

i	C_i	h 1	2	3	4	5	6	F_i
1	82	-	-	-	-	-	-	80
2	91	7	-	-	-	-	-	91
3	93	4	6	-	-	-	-	93
4	94	3	5	7	-	-	-	91
5	92	4	3	4	6	-	-	90
6	104	4	4	5	4	6	-	101
7	103	2	3	5	3	4	6	101

$i = 4$: $C_4 = 94$

$F_4 = 91$

$R_{.4} = 3 + 5 + 7 = 15$

$U_4 = 4 + 4 + 2 + 3 + 4 + 3 + 4 + 5 + 5 = 34$

$V_4 = 6 + 4 + 3 = 13$

$$\hat{a}_4 = \frac{15}{94} = .160 \qquad \text{(eq. 3.7)}$$

$$\hat{M}_4 = \frac{91 \times 34}{13} + 15 = 253 \qquad \text{(eq. 3.8)}$$

$$\hat{N}_4 = \frac{253}{.160} = 1581 \qquad \text{(eq. 3.9)}$$

$$\hat{\sigma}_N{}^2 = 1581(1581 - 94)\left[\frac{253 - 15 + 91}{253}\left(\frac{1}{13} - \frac{1}{91}\right) + \frac{1 - .160}{15}\right] \qquad \text{(eq. 3.10)}$$

$$= 333{,}224$$

$$\hat{\sigma}_N = 577$$

$i = 5$: $C_5 = 92$

$F_5 = 90$

$R_{.5} = 4 + 3 + 4 + 6 = 17$

$U_5 = 4 + 2 + 4 + 3 + 5 + 5 + 4 + 3 = 30$

$V_5 = 6 + 4 = 10$

$$\hat{a}_5 = \frac{17}{92} = .185 \qquad \text{(eq. 3.7)}$$

$$\hat{M}_5 = \frac{90 \times 30}{10} + 17 = 287 \quad \text{(eq. 3.8)}$$

$$\hat{N}_5 = \frac{287}{.185} = 1551 \quad \text{(eq. 3.9)}$$

$$\hat{\sigma}_N{}^2 = 1551(1551 - 92)\left[\frac{287 - 17 + 90}{287}\left(\frac{1}{10} - \frac{1}{90}\right) + \frac{1 - .185}{17}\right] \quad \text{(eq. 3.10)}$$

$$= 360{,}796$$

$$\hat{\sigma}_N = 601$$

$$\hat{s}_4 = \frac{287}{253 - 15 + 91} = .87 \quad \text{(eq. 3.11)}$$

$$\hat{B}_4 = 1551 - 0.87(1581 - 94 + 91) = 178 \quad \text{(eq. 3.12)}$$

NOTE: The variances and standard deviations in this example are large because the numbers of recaptures are small, chosen that way for a simple example.

Estimates of Population Size and Survival when no Recruitment Occurs

If there is no recruitment, all individuals captured after sample i are known to have been alive and present at time i. The number of animals not captured in sample i is $(N_i - C_i)$. Immediately afterwards there are two groups of animals in the population: one is the F_i from which V_i are subsequently caught, and the other is the $(N_i - C_i)$ from which W_i are subsequently caught. W_i is the number of individuals not caught in sample i, both marked and unmarked, but known to be present then because they were subsequently caught and counted ($W_i = U_i$ + the total of unmarked animals caught in each of the samples $i + 1$ through z). Since the probability of capture is assumed to be the same for all animals, $V_i/F_i = W_i/(N_i - C_i)$.

The maximum likelihood estimates (Jolly 1965) are:

$$\hat{N}_i = \frac{F_i W_i}{V_i} + C_i, \quad i = 1, \ldots (z - 1). \quad (3.13)$$

$$\text{Variance of } \hat{N}_i \text{ (for a given } N_i) = (\hat{N}_i - C_i)(\hat{N}_i - C_i + F_i)\left(\frac{1}{V_i} - \frac{1}{F_i}\right). \quad (3.14)$$

$$\hat{s}_i = \frac{\hat{N}_{i+1}}{\hat{N}_i - C_i + F_i}, \quad i = 1, \ldots (z - 2). \quad (3.15)$$

The interpretation of equation (3.15) for the survival rate is obvious.

These formulas are slightly simpler than those required when recruitment occurs, and they also allow estimates for the time of the first sample.

Ex. 3.6. Estimation of population size from multiple mark-recapture data in the special case when no recruitment occurs. The data of example (3.5) are used to illustrate the estimation of N and its variance at $i = 4$ for the case of no recruitment: $C_4 = 94$; $F_4 = 91$; $U_4 = 34$; $V_4 = 13$. The total of unmarked animals alive at $i = 4$ and caught later would normally be acquired by counting the unmarked in catches 5, 6, and 7. In this example the unmarked in each catch equals $(C_i - R_{.i})$; that is, no animal has been recaptured more than once.

$$W_4 = U_4 + \Sigma(C_i - R_{.i}), \text{ for } i = 5, 6, 7.$$

$$= 34 + (92 - 17) + (104 - 23) + (103 - 23) = 270$$

$$\hat{N} = \frac{91 \times 270}{13} + 94 = 1984 \qquad \text{(eq. 3.13)}$$

$$\hat{\sigma}_N^{\,2} = (1984 - 94)(1984 - 94 + 91)\left(\frac{1}{13} - \frac{1}{91}\right) = 246{,}863 \qquad \text{(eq. 3.14)}$$

$$\hat{\sigma}_N = 497$$

A Triple-Catch Method

The multiple mark-recapture procedure requires a minimum of three operations, and such a "triple-catch method" is the simplest way to make mark-recapture estimates when a population is known to have both removal and recruitment. It requires one more operation than does the single mark-recapture, but it is not limited to use in a closed population as are the single and repeated mark-recapture methods. Only two kinds of marks are required: one for those released from the first capture and another for those released from the second (some individuals may be marked twice). The number marked and released from the first capture are the F_1. In the second catch a sample, C_2, is taken, the number previously marked, R_{12}, is recorded, all animals are marked for the second catch, including the R_{12} (or their tag numbers recorded), and F_2 animals are returned to the population. In the third catch a sample of size C_3 is taken and the marked animals are counted according to whether they were previously caught *only* the first time, the R_{13}, or were caught the second time, the R_{23} (including any R_{12} individuals also caught in the third catch). The data are organized as in table 3.2.

The estimates are

$$\hat{M}_2 = \frac{F_2 R_{13}}{R_{23}} + R_{12}, \qquad (3.16)$$

$$\hat{N}_2 = \frac{\hat{M}_2 C_2}{R_{12}} . \tag{3.17}$$

$$\text{Variance of } \hat{N}_2 = \hat{N}_2 (\hat{N}_2 - C_2) \left[\left(\frac{\hat{M}_2 - R_{12} + F_2}{\hat{M}_2} \right) \left(\frac{1}{R_{23}} - \frac{1}{F_2} \right) + \frac{1}{R_{12}} - \frac{1}{C_2} \right] . \tag{3.18}$$

$$\hat{s}_1 = \frac{\hat{M}_2}{F_1} . \tag{3.19}$$

$$\text{Variance of } \hat{s}_1 = \hat{s}_1^{\,2} \left[\frac{(\hat{M}_2 - R_{12})(\hat{M}_2 - R_{12} + F_2)(F_2 - R_{23})}{\hat{M}_2^{\,2} R_{23} F_2} \right] . \tag{3.20}$$

The two variances given above allow only for errors in estimation and do not include the variance due to the stochastic nature of

Table 3.2. Data from the Triple-Catch Method. Symbols are as previously defined.

i	C_i	Recaptures		F_i
1	C_1	-	-	F_1
2	C_2	R_{12}	-	F_2
3	C_3	R_{13}	R_{23}	-

the model. Note that *N* can be estimated only for time 2, *s* only for the interval between times 1 and 2, and that *B* cannot be estimated from data from a triple-catch; the reasons for this can be deduced from examination of the allowable ranges given for equations (3.9), (3.11), and (3.12).

Ex. 3.7. A triple-catch mark-recapture estimate. Hypothetical data for the triple-catch method are tabulated below.

i	C_i	Recaptures h = 1	2	F_i
1	40	-	-	39
2	35	16	-	33
3	31	6	13	-

$$\hat{M}_2 = \frac{33 \times 6}{13} + 16 = 31 \tag{eq. 3.16}$$

$$\hat{N}_2 = \frac{31 \times 35}{16} = 68 \qquad \text{(eq. 3.17)}$$

$$\hat{\sigma}_N^{\,2} = 68(68 - 35) \left[\frac{31 - 16 + 33}{31}\left(\frac{1}{13} - \frac{1}{33}\right) + \frac{1}{16} - \frac{1}{35}\right] \qquad \text{(eq. 3.18)}$$

$$= 238$$

$$\sigma_N = 15.4$$

The number of marked animals in the population at the time of the second catch cannot exceed F_1, but it is possible for $\hat{M}_2$ to exceed F_1 if by chance, or by non-random behavior of the animals, R_{13} is higher or R_{23} is lower than expected on the basis of probability. If $\hat{M}_2$ does exceed F_1, the only safe procedure is to set $\hat{M}_2$ equal to F_1 and calculate $\hat{N}_2 = C_2F_1/R_{12}$, the formula for the single mark-recapture procedure, and to calculate the variance of $\hat{N}_2$ as would be done for that procedure. The survival rate, s_1, would of course equal 1.

An extension of the Jolly-Seber method of estimation from mark-recapture data has been developed by Pollock (1975). It is more general in that it allows for some individual variation in the probability of capture and of survival; at this writing it has not been tested in a practical situation.

3.4 COMMENTS ON MARK-RECAPTURE METHODS

An essential condition for the accuracy of all mark-recapture estimates is that the sampling must be random with respect to each individual's frequency of previous captures. Several tests of "equal catchability" have been proposed; all depend on the frequency of recaptures—that is, whether an individual has been recaptured 0, 1, 2, . . . times after marking—and so require multiple mark-recapture data. Roff (1973b) and Wilbur and Landwehr (1974) review the literature on the subject and find that no one test is reasonably satisfactory. Roff examined nearly 50 applications of the mark-recapture method and found that in most of them the assumption of equal catchability was not supported.

If the basic assumption of the mark-recapture method is rarely true, one naturally questions whether the method is worth using. No other procedure, however, appears to be any more reliable. The solution to the dilemma is to use sampling methods as unbiased as possible and then to avoid attaching much importance to the precision of the estimates.

The relative merits of the three mark-recapture methods may be summarized as follows. The single mark-recapture has the best possibility of making the recapture random with respect to marked and unmarked individuals because the two operations, capture and recapture, can be accomplished by very different methods (see examples in section 1). The requirement of a closed population may be avoided if the estimate is made for the time of first release of marked animals, and if recruits to the population caught in the second capture can be identified (see condition A in the chapter introduction). The repeated mark-recapture probably gives a more precise estimate because of the accumulation of marked animals in the population, but it is weakest with respect to the chances of meeting the basic assumptions, because it is difficult both to ensure a closed population and to randomize the recaptures. Varying the trapping methods is difficult, because animals must be caught and released repeatedly, and about all that can be done is to move traps, change baits, change the direction of net or seine sweeps, or some other minor variations. The multiple mark-recapture allows both removal from and recruitment to the population and also estimates the rates of these. Its disadvantages are that it requires the most complex marking system; it is difficult to randomize recaptures by varying trapping methods; and, because of bias and the correlation between the estimate and its variance, it is hard to evaluate the precision of the estimate.

In a test of several methods on rodent populations of known size, the multiple mark-recapture using Jolly's estimators was more accurate than either the single mark-recapture or the removal method (French *et al.* 1971).

The multiple mark-recapture procedure has an advantage not related to its major purpose of estimating the population, and that is that if the operation is carried on over a relatively long period much information about the population can be accumulated. This may include the lengths of time that individuals remain in the population, distances and direction of dispersal, the proportion that is mature, changes of weight and breeding condition with seasons, and so on. Students of small mammals have used a chart of captures to summarize these kinds of data (see Petrusewicz and Macfadyen 1970, p. 34, for an example). On this chart dates are entered across the top and numbers or symbols identifying individuals down the side. Each capture can be recorded in the appropriate space, including data on weight, condition, and so on. The history of each individual can be read horizontally, and the state of the population at any time can be summarized vertically.

A variation of the single mark-recapture which is practical in its operation, but difficult in its analysis, is to release simultaneously a large number of marked animals, some of which will be recaptured gradually over a long catching period. This procedure is well suited to having the recapture done by a commercial fishery. During the recovery period the number of marked individuals is continually declining because of mortality and the number of unmarked in the population is increasing due to recruitment. A few methods have been developed to estimate the population size at the time of release by allowing for the effects of mortality and recruitment, but some of the assumptions of the models are weak and the calculations are complicated. Seber (1973, chap. 6) describes these methods.

Another method of estimating the size of a population from multiple mark-recapture data is based on the frequency of captures of each individual (Eberhardt 1969, Overton 1971). The data are summarized into the number of individuals caught 1 or 2 or . . . or k times. A frequency distribution is then fitted and extrapolated to estimate the number of individuals, N_0, caught zero times. The population estimate is then the sum of $\hat{N}_0$, and the number of different individuals caught at least once. The method has not given accurate estimates on population of known size (French *et al.* 1971, Phillips and Campbell 1970) apparently because no frequency distribution has accurately estimated N_0.

3.5 METHODS OF MARKING ANIMALS

In a single mark-recapture and the repeated mark-recapture methods all animals can be given the same mark. In the multiple mark-recapture method each individual must be given a separate mark of capture or must be individually identified by a coded mark or numbered tag.

The mark (and the handling during the marking procedure) must not increase the chance of mortality after the animals have been released; it must not interfere with the animals' movements; and it must not affect the probability of marked animals being recaptured. Also the marks must endure. These are rather difficult conditions, and one of the major limitations of the mark-recapture method is the technical problem of marking animals so that all these conditions are fulfilled. Much effort has been made to develop effective methods.

Methods of marking insects are described in detail by Southwood (1966, pp. 57–74).

The greatest variety of marking methods has probably been tried on fish as compared with other animal groups. Fin-clipping is often used but this may affect the survival and mobility of the fish, and these marks can also be lost by regeneration (Ricker 1975, pp. 89–90). Stott (1971) has reviewed the methods of marking and tagging a fish.

Fish too small to be safely marked by the usual means may be marked by dyes (Arnold 1966), but no completely satisfactory dyes have yet been discovered.

Amphibians and reptiles have often been marked by toe-clipping, using a code to mark animals individually. Snakes have been marked by cutting notches in the large ventral scales, and turtles by notching the marginal scutes. As with all methods of marking by mutilation, the survival and behavior of the animals may be changed, and marks may be lost by regeneration; the latter is especially true with amphibians. Turtles can be marked temporarily with quick-drying paint such as fingernail enamel.

The most effective marking methods are available for birds and mammals. Thousands of birds have been individually and safely marked with numbered bands placed around one leg. On the North American continent the banding of wild birds is coordinated by the United States Fish and Wildlife Service (Patuxent Wildlife Research Center, Laurel, Md.), which issues permits, instructions, and numbered bands. Many methods for capturing and marking birds and mammals are described by Taber and Cowan (1971). A variation of the mark-recapture method that has been applied to some birds and mammals is to mark them conspicuously so that the marked individuals can be recognized without being recaptured. The ratio for estimating population size is determined from the number of marked and unmarked animals observed in a counting operation. Taber and Cowan (1971) describe several conspicuous marking methods.

The duration of marks may be tested experimentally by giving two marks, of the same or different kinds, to each individual. Since it is unlikely that both will be lost simultaneously, the recovery of an animal with but one mark surviving can provide data for the rate of loss of the other. Robson and Regier (1966) described such an experiment and concluded that the persistence of tags on fish declined exponentially with time. The loss of leg bands by waterfowl and gulls, estimated in a variety of ways (Hickey 1952, Kadlec and Drury 1968, Ludwig 1967), appears to be low during the first three years after banding and then to accelerate to an appreciable rate during the fourth and following years.

If each animal is given two marks, a correction for loss of marks can be made. In the first case, the two marks are recognizably different and are assumed to have different probabilities of loss. Let one type of mark be designated A, the other B. The number of marked individuals in the recaptured sample will be R_A, R_B, and R_{AB} bearing respectively only mark A, only mark B, or both A and B. The sample will also include an unknown number of individuals which have lost both marks, designated R_0. This number can be estimated as follows. Of those which have kept mark A, the fraction losing B is R_A / R_{AB}; and of those which have lost mark A, the fraction also losing B is R_0 / R_B. If the losses are independent of each other, these fractions are equal and

$$\hat{R}_0 = \frac{R_A R_B}{R_{AB}} \,. \tag{3.21}$$

The actual recaptures are then estimated by $R_{AB} + R_A + R_B + \hat{R}_0$.

If the two tags are identical and have an equal probability of being lost, the same reasoning will lead to the following estimate. Let R_1 be the number of recaptures with one tag, having lost one, and R_2 be the number with two tags; then

$$\hat{R}_0 = \frac{R_1^2}{4R_2} \,. \tag{3.22}$$

Ex. 3.8. Estimating the loss of marks.
Two marks, A and B, are used on each released individual. A sample of the population taken later includes 122 marked individuals, divided as follows (using the symbols given in the text): $R_A = 18$, $R_B = 31$, and $R_{AB} = 73$. From (3.21), the number losing both is $\hat{R}_0 = (18 \times 31)/73 = 7.6$ or 8. The total recapture is 122 + 8 = 130. If the two marks had been of the same kind, the data would be $R_1 = 49$ and $R_2 = 73$. From (3.22), $\hat{R}_0 = 49^2 / 4 \times 73 = 8.2$ or 8.

4 Sex and Age Composition

The proportions of the population that are of each sex and each age describe the population's structure. Not only is this information of descriptive interest, it may tell something of the population's history and may be useful in predicting the reproductive potential and the expected mortality. This chapter is concerned first with how the sex and age composition, separately or together, can be measured and described. It then explains how additional information may be obtained from these characteristics and their changes.

4.1 SEX RATIO

The ratio of the numbers of each sex in the entire population is the *crude sex ratio*, while the ratio of the numbers within a particular age group is a *specific sex ratio*. These ratios are customarily expressed as the number of males per 100 females. As an example of a specific sex ratio, the human sex ratio at birth is close to 106 : 100. The figures may also be expressed as the proportions of each sex within the entire group; thus the proportion of human males at birth is about 51.4 percent, and of females is about 48.6 percent.

Data on sex ratios of many animal species may be found in Spector (1956).

4.2 AGE COMPOSITION

The age of animals may be measured by units of time—such as days, months, or years—or by life history stages even though these may be of unequal duration. Where years are used to measure age, animals living in temperate or arctic climates are conveniently considered to have their "birthdays," or to pass to the next older age group, at a certain time. Fishes customarily are classified as if they were all "born" in midwinter, since growth is the least then. For birds and mammals that reproduce in spring or summer, the most convenient division point is just before the reproductive season. These arbitrary standards bias the statements concerning age, but they simplify classification into age groups. There are several terms used for animals less than one year old (juvenile, calf, etc.); an animal between one and two years is a one-year-old or yearling, after which it becomes a two-year old, and so on.

Insects are often grouped into life history stages such as eggs, nymphal or larval instars, pupae, adults. The same is true for most species which metamorphose. In the same sense, mammals are often divided into juveniles (growing and developing), subadults (full-grown but not reproductive), adults, and occasionally, post-reproductives.

Methods of Determining Age

Methods applicable to aquatic invertebrates are described by Edmondson and Winberg (1971, pp. 149–58). Southwood (1966) has reviewed the means of estimating the age of insects. The taxonomic group with probably the best age-markers is the bony fish, because in temperate and cold climates the alternating growing and non-growing seasons produce yearly annuli in the scales; Tesch (1971) describes this and other methods of determining the age of fish. Several different methods of estimating the ages of game birds, game mammals, and some furbearers are described by Taber (1971). Criteria for estimating age in small mammals are described by Pucek and Lowe (1975).

Some information on age has been obtained by marking individual animals and observing or recovering them later. This is especially true for birds that have been banded as nestlings or as recognizable juveniles.

Interpretation of Age Composition

The age composition of a population in which individuals live through several breeding seasons and the generations overlap, as in many vertebrates, can be represented by an age pyramid as in the examples of figure 4.1. These diagrams may be used to infer something about each population. Assuming that the first two, *A* and *B*, are closed populations of the same species, they can be compared to arrive at the conclusion that the first is increasing relative to the second. The reasoning behind that is that reproductive rates are more likely to vary between populations of the same species than are survival rates, so that the broader base of *A* results from a higher reproductive rate. When we say that the first increases relative to the second, the word *relative* is important, for both populations could be decreasing with *B* shrinking faster. In *C*, on the other hand, the two younger age groups are clearly too small to replace the immediately older age groups and the population is declining. An alternative explanation, if the population were open, is that the bulge in the middle age groups resulted from immigration of these ages.

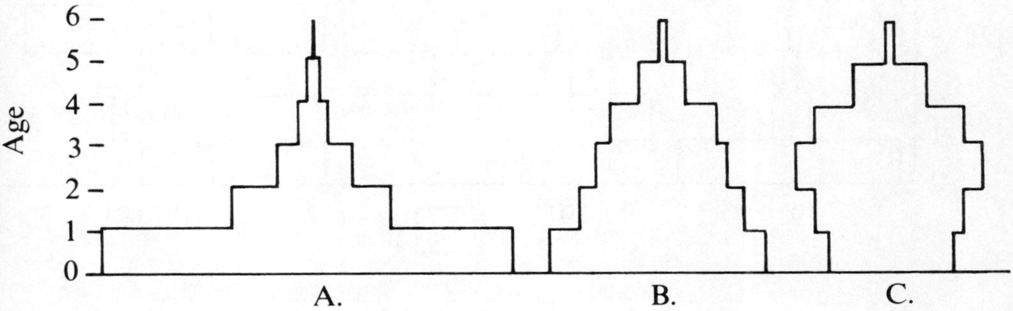

Figure 4.1. Examples of age pyramids. See text for discussion.

If, in addition to the age composition, either the survival rates or the reproductive rates are known, more definite conclusions can be reached about the population's growth. Caughley (1974) illustrates some of the interrelations of these features, and the mathematical relations are described in detail in chapter 9. Figure 9.3 contrasts the age composition of a growing population with the calculated stationary population.

Populations of fishes are likely to have irregular age compositions because spawning success and survival of the very young may change greatly from year to year. In an unusually good year many young are added to the population, and this group may compose a large proportion of the population for the next several years, being known as a "dominant year-class." Figure 4.2 shows the effect of such a year-class on the age composition.

Both the sex and age composition of a population can be illustrated by a pyramid with the proportions of males of each age on one side of the midline and of females on the other, as in figure 4.3. This figure is based on a study (Darling 1951) of two human populations in Scotland, one (A) in a rural area and the other (B) in towns. It is reproduced here to illustrate that age compositions may reveal useful information. For example, in these two populations: 1) the rural population is not replacing itself; 2) the deficiency in the rural population of some groups, especially males between 30 and 60 years

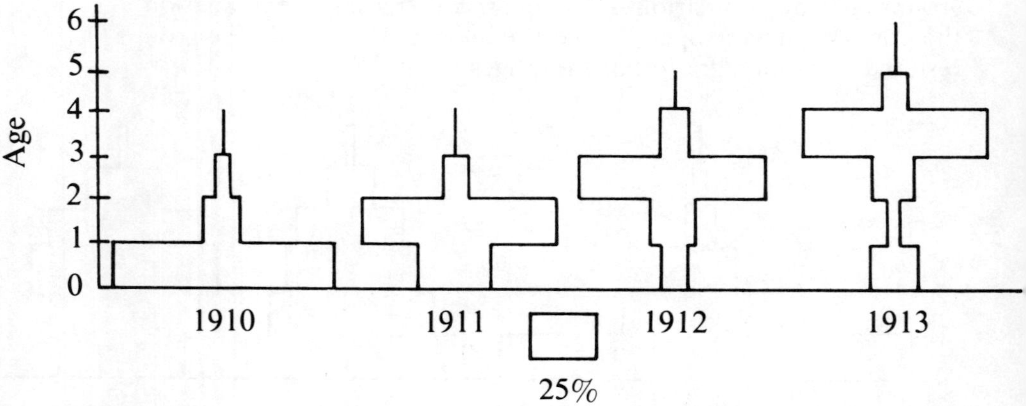

Figure 4.2. Changes in the age composition of a herring population in the North Sea (data from Hjort, 1926). The 1904 year-class was a dominant year-class. The diagrams start with age 6 because the younger ages are too small in body size to be adequately sampled by the fishing gear.

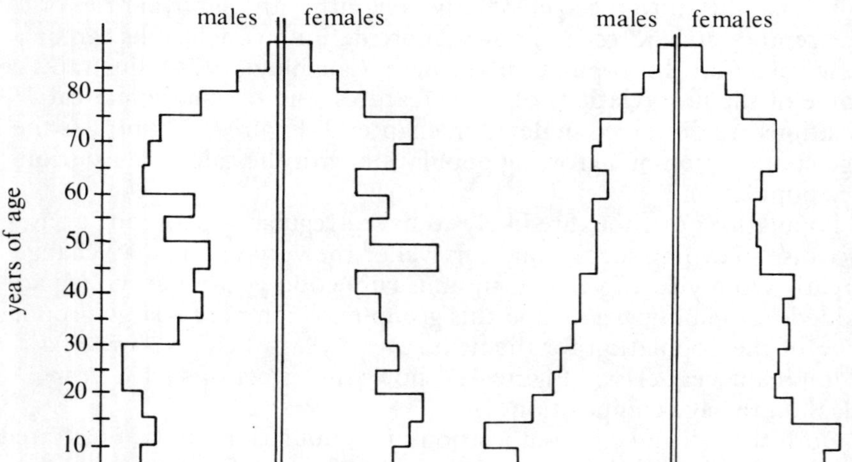

Figure 4.3. Comparison of two human populations in Scotland in 1931 (after Darling 1951). See text for evaluation.

of age, suggests emigration from that area; 3) males predominate in younger age classes because of the unequal ratio at birth; 4) females predominate in older age classes because of their better survival.

Seasonal changes in the numbers of a population, as a whole and in each age group, are shown in figure 4.4A. These are translated into three age pyramids in figure 4.4B to illustrate the effects of reproduction and survival on the age composition at different times of the year.

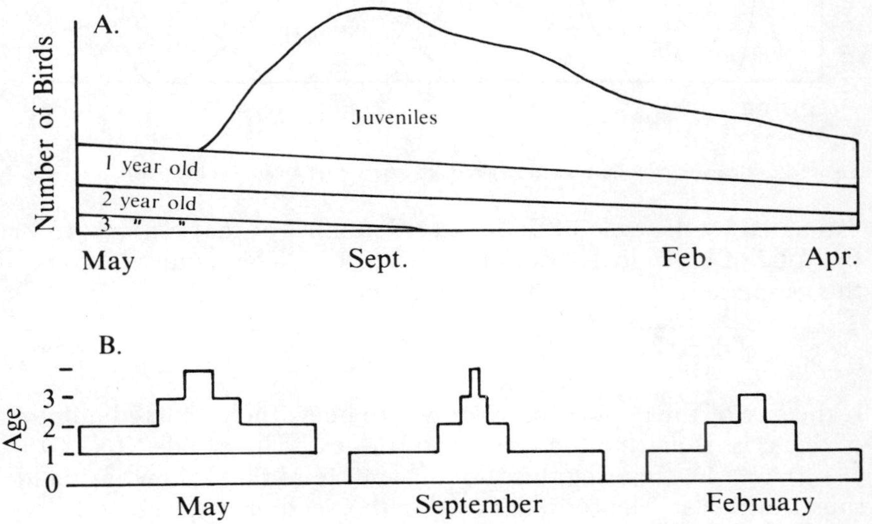

Figure 4.4. Changes in a California quail population (after Emlen 1940). (A) Seasonal changes in numbers. (B) Corresponding changes in age composition.

A species of insect which has an annual life cycle with no overlapping of generations may have changes which can be diagrammed as in figure 4.5. For most of the time such a population consists of just one group or stage, with two being present only when the stages overlap due to some individuals developing and molting faster than others. In species with several generations a year, great overlapping may result by the end of summer from varying speeds of development, and a complex "age" composition may be formed. I know of no examples where such age compositions have been utilized to obtain more information about the population.

4.3 SAMPLING TO ESTIMATE PROPORTIONS

The proportion of any class (e.g., male) in the population is simply

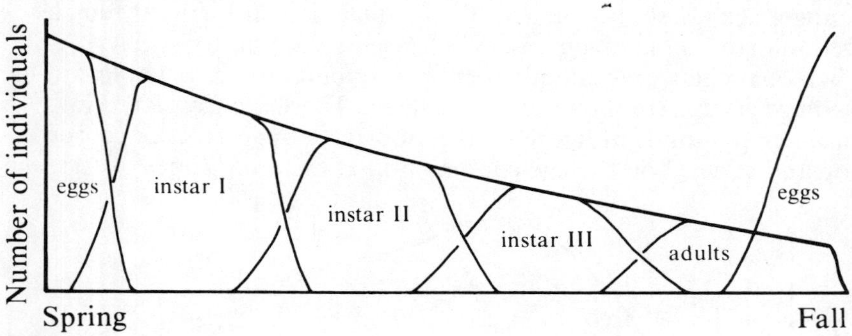

Figure 4.5. Changes in numbers and stages of an insect with an annual life cycle.

estimated by the ratio of the observed number of that class, A, to the total of individuals examined, C; $\hat{p} = A/C$. The standard error of this estimate is

$$\hat{\sigma}_p = \sqrt{\frac{p(1-p)}{C-1}} \tag{4.1}$$

If the size of the population is known to be N, the estimated number of this class is $\hat{p}N$ and the variance of this estimate equals $N(N-C)p(1-p)/(C-1)$. (In using this formula and those that follow, $\hat{p}$ would substitute for p. The formulas used in this section are given with proofs by Cochran [1963, chap. 3], who also includes the finite population correction; the latter can be neglected in most animal studies since usually only a small fraction of the total population is examined.)

Confidence intervals for ratios can be calculated using the normal distribution if the sample size is large, 50 if p—or $(1-p)$, whichever is smaller—is about 0.4 to 0.5, and 200 or more if p is 0.2 or less (Cochran 1963, table 3.3). Preferable to using the normal distribution for this purpose is a table or graph based on the binomial distribution (these can often be found in texts on nonparametric statistics).

The procedures described in the previous paragraphs are like those of simple random sampling, with each individual examined being treated as a sampling unit and classified in one class or another. Frequently, however, ratios are measured in animals by examining all of the individuals in some group, named a "cluster" in sampling terminology. Examples would be all the insects in a sample plot, all the sheep in a flock, and so forth. In this case different calculations should be made. Let there be Q clusters, with C_i individuals in the *i*th cluster of which A_i belong to one class, such as one sex. The ratio is estimated as in the previous case:

$$\hat{p} = \frac{\Sigma A_i}{\Sigma C_i} . \tag{4.2}$$

The estimated standard error is

$$\hat{\sigma}_p = \sqrt{\frac{\Sigma A_i^2 - 2p\Sigma A_i C_i + p^2 \Sigma C_i^2}{Q(Q-1)\overline{C}^2}} , \tag{4.3}$$

where $\overline{C}$ is the average number of individuals per cluster. Example 4.1 applies these two equations.

Ex. 4.1. Calculation of a sex ratio and its variance from cluster sampling. Ten groups of animals were examined and each individual was identified as male or female. The basic calculations were: total examined, $\Sigma C_i = 183$; $\overline{C} = 18.3$; number of males, $\Sigma A_i = 115$; $\Sigma A_i^2 = 1437$; $\Sigma C_i^2 = 3581$; $\Sigma A_i C_i = 2161$. From equations (4.2) and (4.3):

$$\hat{p} = \frac{115}{183} = 0.628$$

$$\hat{\sigma}_p = \sqrt{\frac{1437 - (2 \times .628 \times 2161) + (.628^2 \times 3581)}{10 \times 9 \times 18.3^2}} = 0.0669.$$

If equation (4.1) were used, *p* would be the same but

$$\hat{\sigma}_p = \sqrt{\frac{.628(1 - .628)}{183 - 1}} = 0.0358,$$

which is a biased estimate of the standard error when the data are gathered by cluster sampling.

The above described methods for estimating the ratio of two classes in a population can be extended to three or more classes (Cochran 1963).

4.4 INFORMATION FROM CHANGES IN SEX OR AGE COMPOSITION

Changes in the proportions of the sexes or ages may sometimes be used to obtain additional information about the population. The procedures for doing this are most likely to be useful in game management for at least two reasons: for many game species the differences between the sexes or between juveniles and adults are great enough to allow estimation of the ratios from field observations, and often only males or adults are hunted resulting in distinct changes in sex or age ratios before and after the hunting season. These methods are well reviewed by Paulik and Robson (1969), who term them "change-in-ratio" methods; their paper, or Seber (1973, chap. 9), should be consulted for history, examples, and especially for the discussion of the accuracy of the estimates.

The basic reasoning is outlined in the following example based on a change in sex ratio over a hunting season. The symbols for ratios or quantities which are measured are

C_1: the number of animals observed in the field and identified as to sex at time 1 (before the hunting season).

$\hat{p}_1$: the proportion of the C_1 that are males, an estimate of the proportion of males in the population at time 1. (Note: this is the proportion or fraction, not the sex ratio.)

C_2 and $\hat{p}_2$: the same for time 2 (after the hunting season).

M_d and F_d: the number of males and females, respectively, known to have died between times 1 and 2, known from kill or bag records.

The symbols for quantities which are to be estimated are:

M_1: the number of males in the population at time 1.

F_1 : the number of females, similarly.

$N_1 : M_1 + F_1$.

The actual proportions of males in the population at the two different times are:

$$p_1 = \frac{M_1}{N_1} = \frac{M_1}{M_1 + F_1}\,; \tag{4.4}$$

$$p_2 = \frac{(M_1 - M_d)}{(M_1 - M_d) + (F_1 - F_d)}\,. \tag{4.5}$$

Thus there are two simultaneous equations with two unknowns, M_1 and F_1, and providing that p_1 is unequal to p_2, the equations can be solved for the unknowns and for $N_1 = M_1 + F_1$. The solutions are (replacing p_1 and p_2 by their estimates):

Original number in population:

$$\hat{N}_1 = \frac{M_d - \hat{p}_2(M_d + F_d)}{\hat{p}_1 - \hat{p}_2}\,. \tag{4.6}$$

Original number of males:

$$\hat{M}_1 = \hat{p}_1\hat{N}_1\,. \tag{4.7}$$

Original number of females:

$$\hat{F}_1 = (1 - \hat{p}_1)\,\hat{N}_1\,. \tag{4.8}$$

Chapman (1955) considers in detail the case described in the above example. He gives the large sample variance for the estimates of N_1 and M_1 as follows:

$$\text{Variance of } \hat{N}_1 = \frac{\dfrac{\hat{M}_1\hat{F}_1}{C_1} + \dfrac{(\hat{M}_1 - M_d)(\hat{F}_1 - F_d)}{C_2}}{(\hat{p}_2 - \hat{p}_1)^2}. \tag{4.9}$$

$$\text{Variance of } \hat{M}_1 = \frac{\dfrac{\hat{M}_1\hat{F}_1}{C_1}\hat{p}_2^{\,2} + \dfrac{(\hat{M}_1 - M_d)(\hat{F}_1 - F_d)}{C_2}\hat{p}_1^{\,2}}{(\hat{p}_2 - \hat{p}_1)^2}. \tag{4.10}$$

These equations show that the accuracy of the estimates increases with the sizes of the samples used to estimate the sex proportions and especially with the difference between p_1 and p_2. Chapman, in evaluating this method, concluded that if both the sampling and the removal are done by the research worker, the amount of information yielded by this method is less than for the same effort with mark-recapture procedures. The method will be most useful, therefore, when large numbers of one sex or one age are removed for another purpose, as by hunting.

Ex. 4.2. Estimates of numbers from observed change in sex ratio. Before a hunting season 220 deer (C_i) were observed of which 35 percent ($\hat{p}_1$) were males; similar observations after the season were 190 (C_2) and 26 percent ($\hat{p}_2$). The known kill from hunter checks and field observations was 250 males (M_d) and 10 females (F_d).

$$\hat{N}_1 = \frac{250 - .26(250 + 10)}{.35 - .26} = 2027 \qquad \text{(eq. 4.6)}$$

$$\hat{M}_1 = .35 \times 2027 = 709 \qquad \text{(eq. 4.7)}$$

$$\hat{F}_1 = .65 \times 2027 = 1318 \qquad \text{(eq. 4.8)}$$

$$\hat{\sigma}_N^{\,2} = \frac{\dfrac{709 \times 1318}{220} + \dfrac{(709 - 250)(1318 - 10)}{190}}{(.26 - .35)^2} = 914{,}494 \qquad \text{(eq. 4.9)}$$

$$\hat{\sigma}_N = 956$$

$$\hat{\sigma}_M^{\,2} = \frac{\dfrac{709 \times 1318}{220}.26^2 + \dfrac{(709 - 250)(1318 - 10)}{190}.35^2}{(.26 - .35)^2} = 83{,}236 \qquad \text{(eq. 4.10)}$$

$$\hat{\sigma}_M = 288$$

The general procedure, illustrated above by the example of a change due to differential removal of the sexes, can be applied to changes in age composition, proportion of marked or unmarked animals, and so on, and may involve differential addition to the population instead of removal. The equations for estimating the desired information can be worked out by starting with simultaneous equations similar to (4.4) and (4.5) and solving for the desired quantities. Some of the more useful results are given below ("hats" have been omitted from the estimated ratios to simplify the presentation).

The fraction of animals killed or otherwise removed, for each sex and for both combined, can be calculated from p_1 and p_2 and the sex ratio in those that died. Let $k = M_d/(M_d + F_d)$, the proportion of males in the total number killed; the other symbols are as before.

$$\text{Fraction of males killed: } \frac{M_d}{M_1} = \frac{k(p_1 - p_2)}{p_1(k - p_2)} \tag{4.11}$$

$$\text{Fraction of females killed: } \frac{F_d}{F_1} = \frac{(1 - k)(p_1 - p_2)}{(1 - p_1)(k - p_2)} \tag{4.12}$$

$$\text{Fraction of all animals killed: } \frac{F_d + M_d}{N_1} = \frac{p_1 - p_2}{k - p_2} \tag{4.13}$$

Ex. 4.3. Estimates of hunting mortality from observed change in sex ratio. The data of example (4.2) will be used to estimate the fraction killed.

$$k = \frac{250}{250 + 10} = .96$$

$$\text{Fraction of males killed} = \frac{.96(.35 - .26)}{.35(.96 - .26)} = .35 \qquad \text{(eq. 4.11)}$$

$$\text{Fraction of females killed} = \frac{(1 - .96)(.35 - .26)}{(1 - .35)(.96 - .26)} = .008 \qquad \text{(eq. 4.12)}$$

$$\text{Fraction of population killed} = \frac{.35 - .26}{.96 - .26} = .13 \qquad \text{(eq. 4.13)}$$

Productivity may be defined as the ratio of juveniles, of one or of both sexes, to adult females in the population. In this case, time 1 refers to before the season of recruitment, which is when juveniles mature into the adult age class, and time 2 to afterwards, when all individuals are recognized as adult. New symbols are:

M_j and F_j: numbers of male and female juveniles, respectively, at time 1.

q: proportion of males in the juvenile segment of the population; $q = M_j/M_j + F_j$.

M_1 and F_1: numbers of male and female adults, respectively, at time 1.

M_2 and F_2: the same at time 2, including the newly matured adults.

p_1 and p_2 : the proportions of males in the adult population at times 1 and 2.

The proportions q, p_1 and p_2 are observed in the population. The first may be difficult to obtain because of similar appearances of juvenile males and females, so q is often assumed to be 50 percent if there is no evidence to contradict this assumption.

Productivity as measured by juvenile females per adult female,

$$\frac{F_j}{F_1} = \frac{(p_1 - p_2)(1 - q)}{(1 - p_1)(p_2 - q)}. \qquad (4.14)$$

Productivity as measured by all juveniles per adult female:

$$\frac{F_j + M_j}{F_1} = \frac{p_1 - p_2}{(1 - p_1)(p_2 - q)}. \qquad (4.15)$$

Ex. 4.4. Estimates of productivity from observed change in sex ratio. Counts in early summer of adult ring-necked pheasants estimated the proportion of males (p_1) to be .25. In early fall (birds of the year will be in adult plumage), the proportion of males (p_2) was .37. The sex ratio of juveniles is close to 100 : 100 or $q = .5$.

$$\text{Juvenile females per adult female} = \frac{(.25 - .37)(1 - .5)}{(1 - .25)(.37 - .5)} = .62 \qquad \text{(eq. 4.14)}$$

$$\text{All juveniles per adult female} = \frac{(.25 - .37)}{(1 - .24)(.37 - .5)} = 1.21 \qquad \text{(eq. 4.15)}$$

Relative survival rates of males and females can be easily estimated from changes in the proportions of the sexes. Defining the survival rate of males over the period from time 1 to time 2 as $s_M = M_2/M_1$ (that is, the fraction that survives), and doing the same for females, it is easy to use equations (4.4) and (4.5) to show that

$$\frac{s_M}{s_F} = \frac{p_2 (1 - p_1)}{p_1 (1 - p_2)}. \qquad (4.16)$$

Paulik and Robson (1969) discuss several aspects of these methods that have often been neglected; these include the estimation of the effects of bias, the estimation of variances, and the calculation of confidence intervals. They present charts to aid in determining the sample sizes required for estimations within a certain error, and they illustrate the methods with numerous examples. Seber (1973) also reviews the literature and the mathematics on which these estimates are based.

5 Introduction to Population Rates

Population numbers change primarily because of births and deaths and secondarily because some individuals migrate into or out of the population. The frequency with which each of these events occurs is measured by a population rate, such as the birth or death rate. The next three chapters will discuss separately several rates which are important in ecological research and in understanding population dynamics. But first it is necessary to comprehend the meaning and general properties of any population rate; these are explained in a step-by-step process in this chapter.

To begin with an example. If D individuals of a population die during a year, the death rate could be defined as D/year. This is a measure of the frequency of deaths within that population, and is known as the *absolute death rate*. It is not, however, useful for comparison with another population, or the same at another time, unless the population sizes are equal. A more useful expression is obtained by dividing the absolute death rate by the number in the population, N. The resulting ratio, written either as D/N/year or $D/N \cdot$ year is the *population death rate*. When the shorter phrases, "death rate," "mortality rate," and so forth are used, almost invariably is meant a population rate.

If a population rate is measured with respect to the entire population, it is a *general population rate*. If it is measured with respect to some specified part of the population, it is a *specific population rate*. An example of the latter is the annual number of deaths of males aged 0 to 6 inclusive per all males of that age class in the population; this is an age-sex specific death rate. (Note: The general population rate defined above is customarily called a "crude" rate; "general" seems, however, a better antonym to "specific" than does "crude.")

The observed ratio of deaths to population size is usually adjusted to something like deaths per 1000; a larger base (e.g., 10,000) would be used if more accuracy is warranted and in order to avoid using fractional numbers of deaths. A rate may also be expressed as a decimal. Examples of how the same population death rate may be stated are: m = 509/10,000/year = 509/10,000 person-years = .0509/year.

The different ways of measuring this rate will be illustrated by the example of figure 5.1. At the start of an interval the original number

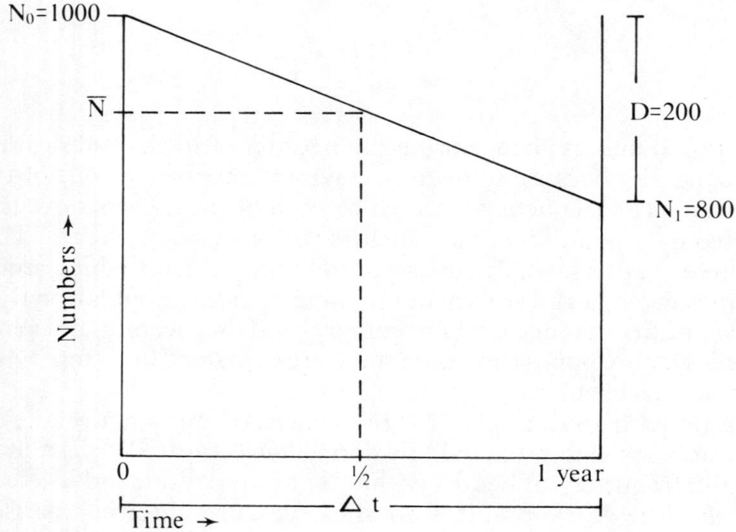

Figure 5.1. Symbols used in measuring a death rate.

of the population N_0 = 1000. During the interval, one year in the example, the number dying D = 200. It is assumed that no other changes occur, so the number at the end of the interval N_1 = 800. One way to measure the mortality is in terms of the number alive at the start of the interval,

$$m = \frac{D}{N_0 \Delta t} . \qquad (5.1)$$

In this example m = 200/1000 person-year = .200/year.

Only at the start of the year, however, are 1000 individuals exposed to the chance of death and near the end of the year only 800 are so exposed. This suggests the concept of defining the rate in terms of the average number alive during the interval,

$$d = \frac{D}{\overline{N} \Delta t} . \qquad (5.2)$$

If the only information we have is that given in the above example, the usual way of estimating the average number is $\overline{N} = \frac{1}{2}(N_0 + N_1)$. Using equation (5.2), d = 200/900 person-years = .222/year.

From now on these two measures will have different names. The first, where the population base is N_0, is the mortality rate, m; it is also the probability of death in the interval since it is the number dying divided by the number that could die. The second, using $\overline{N}$, is the death rate, d. This latter is the measure most often used in demography.

The differences between the two ways of measuring mortality may be reduced by using a shorter time interval. As Δt approaches zero as a limit, the two measures converge because $\overline{N}$ will approach N_0 as a limit. It is convenient here to change the notation, replacing D with $-\Delta N$, the decrease in numbers due to death. Then as Δt approaches zero, N_0 approaches $\overline{N} = N$ and calculus notation may be used:

$$\Delta t \rightarrow 0,\ m = d = \frac{1}{N}\left(-\frac{\delta N}{\delta t}\right),$$

or

$$\frac{\delta N}{\delta t} = N(-d). \tag{5.3}$$

This may be called an instantaneous population rate implying that the changes are continuous. Actually populations change by discrete jumps as whole individuals are added or removed, and this fact has raised questions about the appropriateness of instantaneous rates. But L. C. Cole (1954) carefully examined this inconsistency, and he concluded that, except when numbers are very small, instantaneous rates can be used to measure population changes. This is useful because it permits the use of the powerful methods of calculus.

Equation (5.3) can be solved for the number remaining at any time t:

$$N_t = N_0 e^{-dt}, \tag{5.4}$$

where e is the base of natural logarithms. This expression says that, under the conditions of this example, the population declines along an exponential curve like that of figure 5.2.

Equation (5.4) can be used to calculate an instantaneous death rate if N_0 and N_t are known. From the example used above, N_1 = 800 = $1000^{-d(1\text{ year})}$; solving, d = (ln 1000 - ln 800)/year = .223/year. This equation illustrates the point that, even though the death rate is an instantaneous rate, it must be stated as a number per unit of time.

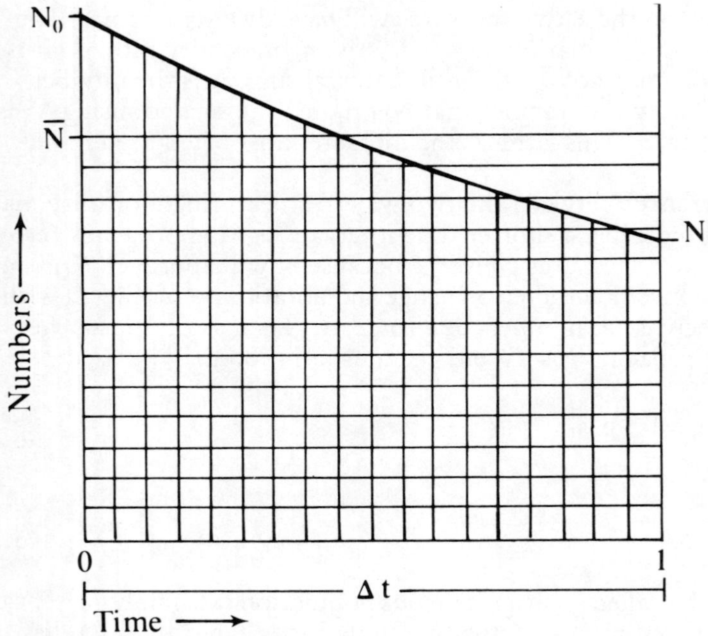

Figure 5.2. Exponential decline in numbers when death rate is constant. See text for discussion concerning $\overline{N}$.

Both calculus and intuition can lead to a useful concept of the average number alive, $\overline{N}$, and of the death rate as follows. In figure 5.2 two areas are indicated. One is marked by horizontal hatching and is the rectangular area equal to $\overline{N}\Delta t$. The other is the area shown by vertical hatching that lies underneath the exponential curve; this, from equation (5.4), is equal to the definite integral

$$\int_0^1 N_0 e^{-dt} \delta t,$$

which equals $\Delta N/d$. If $\overline{N}$ is chosen so that these two areas are equal, we have $\overline{N}\Delta t = \Delta N/d$. Solving for d gives $d = \Delta N/\overline{N}\Delta t$, as in equation (5.2). Therefore we have the useful relation that the instantaneous death rate defined in equations (5.3) and (5.4) equals the death rate based on the average number *if* $\bar{N}$ is calculated so that the area under it equals the area under the curve of population decline. Study of figure 5.2 should also show intuitively that $\bar{N}$ as defined in this paragraph is also the mean value of N_t averaged over each instant of Δt. Returning to consideration of the area under the curve, we can see that this area is the summation of the time lived by all individuals surviving all or part of the period Δt; it is the total time lived by the population members during which death could have occurred. This leads to the concept of a population death rate being the ratio between the number of deaths in a specified period and the total exposure within that period to the risk of death.

Since measurement of the total time lived by the population or of the average number of individuals alive in an interval is difficult in practice, $\bar{N}$ is usually approximated from $\frac{1}{2}(N_0 + N_1)$, which is the midpoint of the straight line connecting N_0 and N_1 in figure 5.1. This estimate is satisfactory if Δt is short. This is also an estimate of the population number at the midpoint of the interval, and a frequent practice in demography is to census the population at the midpoint and to use the result as $\bar{N}$. The methods of estimating rates for human populations are described in detail by Barclay (1958) and Spiegelman (1968).

Inclusion of the time period, Δt, in the ratio can lead to some awkward expressions. So it is customary to omit it from the ratio and instead to state the interval in the name of the rate (e.g., the "annual death rate" or the "daily growth rate"). This will be the practice in the following chapters unless the time period needs to be included for mathematical consistency.

6 Mortality and Survival

Much ecology is "a matter of life and death." Research on the suitability of the environment and the fitness of the population often measures these by the frequency of deaths, or to look at the other side of the coin, by the chances of survival. Applied ecologists wish to separate the effects of natural mortality from those of, for example, mortality by hunting or fishing. Furthermore, age-specific death rates have a great impact on the age structure of the population and its potential reproductive rate. It is not surprising, therefore, to find that much work has been done on the statistics of death and survival, so much in fact that the subject can be confusing to a student trying to find the measurements that he needs. This chapter describes first the meaning and measurement of general rates, then the construction of a life table and the measurement of age-specific rates, statistical tests that may be applied, and the significance of longevity and turnover. In the final section is a summary and guide to the methods which is organized according to the nature of the data.

6.1 MORTALITY, DEATH, AND SURVIVAL RATES

Mortality rate is defined as the probability of death in an interval of time (symbols as in previous chapter; see fig. 5.1): $m = D/N_0$. By definition, it applies only to the N_0 individuals alive at the beginning of the period.

Death rate is based on the average population size or on the summation of time lived in the interval during which the deaths are counted (see preceding chapter for a discussion of this); $d = D/\overline{N}$. In a general death rate all deaths are counted including any of the individuals born during the interval; the newborn are also included in the calculation of the average population size.

The survival rate is the probability of survival, and therefore is the complement of the mortality rate: $s = N_1/N_0 = 1 - m$.

Survival rates have advantages over mortality rates (besides having a more cheerful meaning) when certain calculations are required. One of these is that survival rates for successive periods may be multiplied to calculate the survival rate for a longer period. For example, if s_1, s_2, and s_3 are the survival rates for each of three successive months, the multiple product $s_1 s_2 s_3$ is the survival rate over the three month interval (this is easily proved by considering these rates as probabilities, or by writing each according to the definition of a survival rate and then multiplying). Conversely, if for example the annual survival rate is known and if it is assumed that the monthly survival rates are equal,

$$s_{month} = \sqrt[12]{s_{year}}.$$

Another reason for emphasizing survival rate is that it, more often than mortality rate, can be calculated directly from the data.

For purposes of summary and comparison, presented below are definitions of general mortality, survival, and death rates, and equations relating these rates. See figure 5.1 for illustrations of the symbols. The intervals used in measuring these rates are all equal.

A. Mortality rate:

$$m = \frac{D}{N_0}. \qquad (6.1)$$

B. Survival rate:

$$s = \frac{N_1}{N_0} = 1 - m. \qquad (6.2)$$

C. Death rate:

$$d = \frac{D}{\overline{N}}. \qquad (6.3)$$

D. Under either of two conditions, mortality and death rates are equal (i.e., $m = d$):

1. when the interval Δt approaches zero as a limit (see equation 5.3);
2. when losses by death are continually replaced by reproduction so that N_0 equals $\overline{N}$.

E. When mortality alone affects the numbers (no migration or reproduction) and the rates remain constant throughout the period, the rates are interrelated as follows:
   1. If the average number, $\overline{N}$, is based on the total time lived in the interval (see discussion concerning fig. 5.2), equations (5.4) and (6.3) can be used to prove that:

$$s = e^{-d}, \text{ and } m = 1 - e^{-d}. \tag{6.4}$$

   2. If $\overline{N}$ is approximated by $(N_0 + N_1)/2$, algebraic manipulation of (6.1) and (6.3) and this approximation of N shows that

$$m = \frac{2d}{2 + d}, \quad d = \frac{2m}{2 - m}. \tag{6.5}$$

Ex. 6.1. Calculation of general mortality, survival, and death rates. The numbers are from figure 5.1, which assumes that there is neither reproduction nor migration.

$$m = \frac{200}{1000} = .200 \qquad \text{(eq. 6.1)}$$

$$s = \frac{800}{1000} = 1 - .200 = .800 \qquad \text{(eq. 6.2)}$$

Estimating $\overline{N}$ by ½(1000 + 800) = 900:

$$d = \frac{200}{900} = .222 \qquad \text{(eq. 6.3)}$$

$$m = \frac{2 \times .222}{2 + .222} = .200 \qquad \text{(eq. 6.5)}$$

Basing $\overline{N}$ on the total time lived (area under the curve of figure 5.2):

$$d = \ln \frac{1000}{800} = .223 \qquad \text{(eq. 5.4)}$$

$$s = e^{-.223} = .800 \qquad \text{(eq. 6.4)}$$

To illustrate projection through a second Δt to time $t = 2$:

$$N_2 = 800s = 800 \times .800 = 640$$

or

$$N_2 = 800e^{-.223} = 640$$

Figure 6.1 graphs the relative values of m and d for the conditions and equations given above; note that if mortality alone is changing the population, *m* equals or is less than *d*, and equation (6.5) is a good approximation to (6.4).

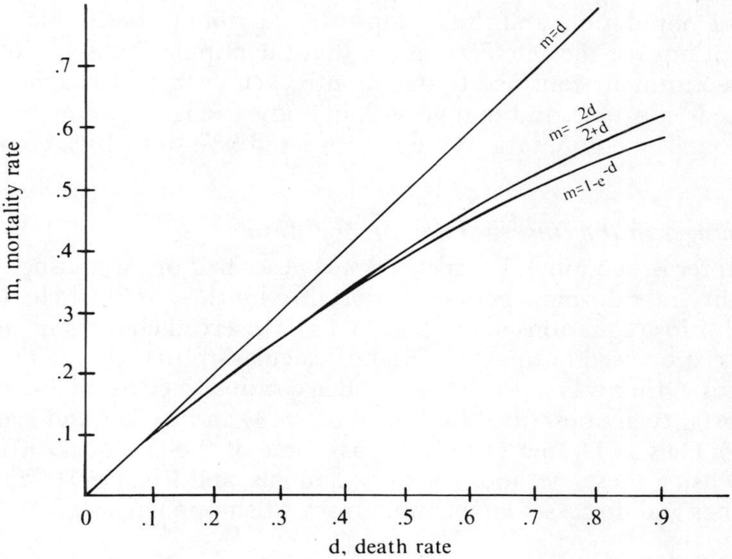

Figure 6.1. Equivalent values of the mortality rate, *m*, and the death rate, *d*. The curved lines apply to the case of a population with no migration or reproduction, only mortality.

6.2 ESTIMATION OF GENERAL RATES

Using Population Records

In a human population the data for computing the annual general (or crude) rates are obtained from the number of deaths registered in a year and the population size either at the start of the year (for mortality rate) or at mid-year (for death rate). Barclay (1958, sec. 2.11) describes the conventional methods of estimating the population size when the census data are acquired at a time other than the start or the middle of the year.

If a population is closed (lacking migration) and the numbers at the start and end of an interval, N_0 and N_1, and also the number added by reproduction, B, are all known, then the number dying in the interval is $D = N_0 + B - N_1$. Southwood (1966, p. 247) cites several studies of insects in which mortality has been measured by this principle.

A method for estimating both the mortality rate and the growth in body size of individuals in a population has been described by Ebert (1973). The required data are: average body size in the population at two times during the year, average size of individuals at recruitment into the population, and the asymptotic (maximum) body size. The assumptions for the calculations are that the population is stationary with recruitment confined to one month each year, that the mortality rate is constant, and that growth in body size is an exponential function. The computations are iterative and best done by a computer.

By Changes in the Observed Sex or Age Ratios

In chapter 4, section 4.3, a method was described of estimating the mortality rate during a period of mortality by the observed changes in sex ratios (equations 4.11, 4.12, 4.13). Observed changes in age ratios can be used in the same kind of calculation to estimate mortality rates. Other ways of using sex and age ratios to elicit information on mortality are described by Hanson (1963) and Kelker and Hanson (1964). Geis and Taber (1963) discuss some of the practical difficulties of using these methods for game animals, and Ricker (1975) describes and discusses an example from a fish population.

Using Mark-Recapture Methods

In Chapter 3 was described the multiple mark-recapture method of estimating population parameters, including survival. Equations for the survival rate are (3.11), (3.15), and (3.19). Any rates so calculated apply only to the segment of the population that can be marked and recaptured; this restriction applies also to each of the following methods.

The other methods that depend upon the recapture of previously marked animals use the ratio of recoveries to marked in population as an estimate of the probability of capture. One such method requires one mark and two recaptures. The number recaptured at time 1 will be the probability of being captured, $P(C)$, times the number of marked animals, M_1, in the population at that time: $R_1 = P(C)M_1$. At time 2 the number of marked animals will be M_1s, where s is the survival rate over the interval, and if the catch effort and other conditions affecting catch success are the same (an important but often incorrect assumption), $R_2 = P(C)M_1s$. Dividing the second by the first equation:

$$s = \frac{R_2}{R_1}. \qquad (6.6)$$

A second method avoids the assumption made above that the probability of capture is equal at two different times. It requires two mark and release operations, with different marks being used each time, and one recapture, at which time the probability of recapture is estimated. The steps of the procedure are as follows:

1. At time 1, M_1 individuals are marked and released.
2. Shortly before time 2, a period later, M_2 individuals are marked differently and released.
3. Very soon after this second release a recapture is made which will obtain R_1 individuals from the first marking and R_2 from the second. On the assumption that the M_2 animals are still present, the probability of capture is estimated by R_2/M_2.
4. The number of recaptures from the first marking equals the probability of capture times sM_1, the number of survivors, or the estimated survival rate for the period 1 to 2 is $\hat{s} = R_1M_2/R_2M_1$.

Seber (1973, pp. 222–23) gives the following as the approximately unbiased estimates of the survival from time 1 to 2 and its variance:

$$\hat{s} = \frac{R_1(M_2 + 1)}{(R_2 + 1)M_1}, \tag{6.7}$$

$$\text{Variance of } \hat{s} = \hat{s}^2 - \frac{R_1(R_1 - 1)(M_2 + 1)(M_2 + 2)}{M_1(M_1 - 1)(R_2 + 1)(R_2 + 2)}. \tag{6.8}$$

Ex. 6.2. Estimating survival by a mark-recapture method. Three hundred fish were marked and released on June 1. On September 25, 75 were marked differently and released. On October 1, 250 fish were caught, of which 35 bore the June mark and 12 bore the September mark.

$$\hat{s} = \frac{35(75 + 1)}{300(12 + 1)} = .68 \text{ survival over 4 months} \qquad \text{(eq. 6.7)}$$

$$\hat{\delta}_s^{\,2} = .68^2 - \frac{35(35 - 1)(75 + 1)(75 + 2)}{300(300 - 1)(12 + 1)(12 + 2)} = .03583 \qquad \text{(eq. 6.8)}$$

$$\hat{\delta}_s = .19$$

Ricker (1975, Chapter 5) describes and gives examples of this method as applied to fisheries. Geis and Taber (1963, pp. 285–87) describe it as applied to game animals, and give an example of how it can be continued for several years. Seber (1973, pp. 287–95) develops the general theory behind the method and includes variations in its use and analysis. A limitation of this method is that the probability of capture must be estimated immediately after the second (or later) marking, making it most useful for game animals killed or fish caught in a short open season; these operations are usually not random with respect to age so the results are applicable only to a segment of the population. The first two works cited above discuss some of the practical problems in collecting and evaluating data needed for the estimation of survival rates.

Another source of information which can be used to estimate survival is the number of animals found dead in a multiple mark and release study. The major assumption of this method (Seber 1970, 1973) is that the survival rate and the probability of dead individuals being recovered are functions only of the year (or other period) in which the survival or death occurs and not of the age of the individuals; a survival rate independent of age is reasonable for some species (see section 6.5).

The data required and the symbols for each (changed from those of Seber) are described below and are illustrated in figure 6.2. This figure, which is a modification of figure 6.6 to be explained in section 6, is a useful way to organize the data.

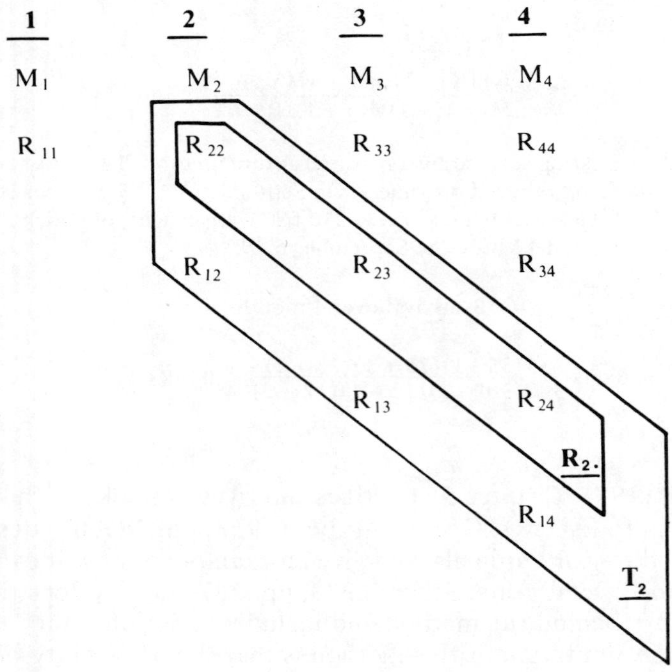

Figure 6.2. Explanation of the symbols and organization of the data for the estimation of an annual survival rate from the recovery of dead, marked animals. Each diagonal row of R_{ij} extending downward and to the right is the record of recoveries from one year's marking. The boxes enclose examples of two sums, $R_2.$ and T_2.

M_i : number marked and released at start of year. The year i begins with the release of M_i individuals, with i ranging from 1 to w.

R_{ij}: number from the M_i individuals that were recovered dead within year j; for example, year 3 begins with the release of M_3 individuals, and all animals recovered dead within that year are counted in one or another of R_{13}, R_{23}, or R_{33}.

The symbols for certain calculations are as follows:

For all individuals marked at time i and recovered within that year or later,

$$R_{i\cdot} = \Sigma_{j=1}^{w} R_{ij}\,.$$

For those marked at time $(i + 1)$ and recovered later, $R_{i+1\cdot}$ is obtained by a similar sum.

For the total number marked at time i or earlier and recovered after time i,

$$T_i = \Sigma_{h=1}^{i} \Sigma_{j=1}^{w} R_{hj}\,.$$

For the total marked at time $(i + 1)$ or earlier and recovered later, $T_{i+1\cdot}$ is similarly calculated.

Seber (1970) gives the maximum likelihood estimate of the survival rate for year i as:

$$\hat{s}_i = \frac{(T_{i+1} - R_{i+1\cdot})\, M_{i+1}\, R_{i\cdot}}{T_i\, M_i\, R_{i+1\cdot}}, \tag{6.9}$$

and the asymptotic variance of the rate as (omitting expected value symbols):

$$\text{Variance of } \hat{s} = \hat{s}^2 \left(\frac{1}{R_{i\cdot}} + \frac{1}{R_{i+1\cdot}} + \frac{1}{T_{i+1} - R_{i+1\cdot}} - \frac{1}{M_{i+1}} - \frac{1}{T_i}\right). \tag{6.10}$$

An explanation of (6.9) is fairly simple. The estimation of the survival rate is based on estimations of the total number marked and known to be alive at time i, the N_i, and of the survivors of the N_i alive at time $(i + 1)$. The N_i can be estimated because the most likely value of T_i, the number of marked animals alive at time i and recovered later, is N_i multiplied by the probability of recovery after time i, or $T_i = N_i(R_i/M_i)$. At time $(i + 1)$ there are present the survivors of the N_i plus those newly marked, totalling $(sN_i + M_{i+1})$; the expected value of T_{i+1} is this total multiplied by the probability of recovery after $(i + 1)$, or $T_{i+1} = (sN_i + M_{i+1})(R_{i+1}/M_{i+1})$. Manipulation of these two equations will yield (6.9).

An important source of error in all estimates of survival rate from mark-recapture data is that arising from the loss of tags or marks. This problem is discussed in section 6.6 of this chapter under the subheading "Using Marked Individuals to Estimate Rates."

Ex. 6.3. Estimation of survival rate from the recoveries of dead, marked animals. The data from a hypothetical example of the recovery of marked dead animals, marked and recovered over four years, are tabulated below as in figure 6.2. The s_2, survival from year 2 to 3, is estimated.

	year			
	1	2	3	4
M_i	210	224	193	206
R_{ij}	52	67	42	54
		38	27	33
			15	21
				12

To estimate s_2:

$R_{2\cdot} = 67 + 27 + 21 = 115;$
$R_{3\cdot} = 42 + 33 = 75;$
$T_2 = 67 + 27 + 21 + 38 + 15 + 12 = 180;$
$T_3 = 42 + 33 + 27 + 21 + 15 + 12 = 150;$

$$\hat{s}_2 = \frac{(150 - 75) \times 193 \times 115}{180 \times 224 \times 75} = .55 \qquad \text{(eq. 6.9)}$$

$$\hat{\sigma}_s^{\,2} = .55^2 \left(\frac{1}{115} + \frac{1}{75} + \frac{1}{150 - 75} - \frac{1}{224} - \frac{1}{193} - \frac{1}{180}\right) = .0061$$

$$\hat{\sigma}_s = .078$$

Mortality Rates by Cause of Death

The methods for measuring the effects of different causes of mortality have largely been developed by fishery biologists who wish to measure population losses due to fishing and those due to other or natural causes. The death rate by cause can be defined as follows. For symbols, let $D.$ be the total number of individuals dying in a period of time; let the sum of those dying from cause f be D_f; and let the sum of those dying from cause g be D_g. The death rates by each cause are $d_f = D_f/\overline{N}$ and $d_g = D_g/\overline{N}$, and from both causes is $d. = d_f + d_g$.

Note that these are not specific death rates, as defined in the first part of chapter 5, because they are measured with respect to the entire population. Before discussing methods of measuring death rates by cause, it will be useful to describe the relations between death, mortality, and survival rates by cause.

The general survival rate and general mortality rate are related as usual: $s = 1 - D./N_0 = 1 - m$. Define s_f as the probability of surviving given that f is the only cause of death, and m_f, the probability of dying under this condition; then $m_f = 1 - s_f$. If only cause g is effective, s_g and m_g are defined similarly. If both causes are operating, and if each individual is vulnerable to death by either, the general survival and mortality rates are:

$$s. = s_f s_g$$

and

$$m. = 1 - s_f s_g = m_f + m_g - m_f m_g. \tag{6.11}$$

Morris (1965) has made equation (6.11) more general by including an "index of vulnerability," symbolized by v: $m. = m_f + m_g - v m_f m_g$. If the two causes of death are exclusive—animals affected by one are not affected by the other—then v equals zero. If the entire population is vulnerable to both, v equals one. If animals affected by one factor are more vulnerable to the other, v exceeds one; an example is that heavily parasitized animals are more vulnerable to pneumonia than are the unparasitized.

The relation between death rate and survival rate by cause can be simply expressed if the death rate acts continuously through the period; from equations (6.4),

$$s_f = e^{-d_f},$$

$$s_g = e^{-d_g},$$

$$s. = e^{-(d_f + d_g)}. \tag{6.12}$$

One method of estimating mortality by cause depends upon the use of marking and releasing individuals. Its application in fishery biology is described by Beverton and Holt (1957), Cormack (1968), and Cushing (1968), and it will be illustrated here in terms of losses due to fishing and losses due to other causes. Let d_f be the death rate in the population due to fishing, and d_g include losses from other causes. The procedure begins with the marking and releasing of fish (the more the better) at the beginning of a fishing period. Marked fish caught in the harvest are counted. A second period follows, of the same length and fishing intensity, and the marked fish in this catch are counted. The data are:

M: number of fish marked and released at time 0, start of period 1.

R_1 and R_2: number of marked fish caught in periods 1 and 2, respectively.

T: the length of each fishing period.

The marked fish are assumed to have the same vulnerability to death from either cause as the unmarked. If this is true, and if the fishing intensity in the two periods is equal, the general survival rate (equation 6.12) and its estimate (6.6) are for a period T:

$$s = e^{-(d_f + d_g)T} = \frac{R_2}{R_1}. \qquad (6.13)$$

The probability of death from fishing in the first period and its estimate is $D_f/N_0 = R_1/M$. This is also the probability of dying for any reason times the probability that a death is caused by fishing, or $D_f/N_0 = m.d_f/(d_f + d_g)$. Combining these and equation (6.13), it can be shown that the death rate from fishing over one period can be estimated from

$$\hat{d}_f = \frac{R_1 \ln(R_1/R_2)}{MT(1 - R_2/R_1)}. \qquad (6.14)$$

All quantities on the right side of this equation are known, so d_f can be calculated. Using this result, d_g can be calculated from (6.13).

Ex. 6.4. Estimation of death rates from fishing and from natural causes. Eight hundred fish were marked and released at the start of year 1. Of these, 121 were caught in the first year and 55 in the second. The fishing intensity was judged to be the same in both years.

$$d_f = \frac{121 \ln(121/55)}{800(1 \text{ year})(1 - 55/121)} = .22/\text{year} \qquad (\text{eq. } 6.14)$$

$$e^{-(.22/\text{year} + d_g)(1 \text{ year})} = \frac{55}{121} \qquad (\text{eq. } 6.13)$$

$$d_g = .57/\text{year}$$

The above method can be extended to using data from more than two fishing periods. Another method of separating fishing and natural mortality is to use variations in fishing effort or intensity and the resulting variations in general survival with the assumption that natural mortality remains constant. Descriptions of these and other methods can be found in Beverton and Holt (1957), Cormack (1968), Cushing (1968), Ricker (1975), and Seber (1973).

6.3 AGE-SPECIFIC RATES AND THE LIFE TABLE

An age-specific mortality rate is the ratio of the number of deaths of individuals between the ages of a and $(a + 1)$ to the number alive at exact age a. It is the probability that an individual attaining age a will die within the next year, and its complement, the age-specific survival rate, is the probability that it will survive the year. The study of these rates is especially important to actuaries, and they long ago developed the device of the life table for this purpose. The history and importance of the life table for the study of human mortality and longevity are described by Dublin and Lotka (1936). The simplest way to understand the life table is to begin by imagining a group of individuals all born simultaneously; this group is called a *cohort*. It may contain only one sex, as is usual in a life table for humans, or it may contain both sexes. The life table presents the history of this cohort by tabulating the number of survivors remaining at regular intervals such as a year, until the last individual is dead. From the survivorship data can be calculated several characteristics of the cohort. One life table will be described in some detail.

The symbols used for the column headings of table 6.1 have been changed from the conventional ones to be consistent with the symbols used in the rest of this book; the conventional symbols follow in parenthesis.

Headings of life table columns (table 6.1):

a (x) : age at beginning of interval, number of whole intervals lived.

L_a (l_x) : number of survivors of the cohort at exact age a.

D_a (d_x): number of individuals dying between ages a and $(a + 1)$.

m_a (q_x): probability of the L_a individuals dying before age $(a + 1)$, the age-specific mortality rate.

$\overline{L}_a$ (L_x): average number of survivors between ages a and $(a + 1)$.

$\bar{e}_a$ (e_x) : expectation of further life at exact age a.

The L_a column is the survivorship column. The D_a column is the record of deaths. Obviously this and the preceding column are related by $L_a - D_a = L_{a+1}$. Also, the sum of all $D_a = L_0$. The age-specific mortality rate in the next column is $m_a = D_a / L_a$. In some life tables this column is expressed in the form $1000m_a$ to avoid using a decimal point. This column and the preceding two present the same information, because if one of them is known, the other two can be calculated.

The numbers in the column $\overline{L}_a$ may be thought of as (1) the average number of the cohort alive during the interval, or as (2) the total number of years lived by members of the cohort between exact ages a and $(a + 1)$. In most cases it is approximated by $\overline{L}_a = (L_a + L_{a+1})/2$. The exception is the first year, when deaths are more likely soon after birth and then the probability of death decreases. For table 6.1 $\overline{L}_0$

was estimated by assuming a high mortality at first which diminished until at the end of the first year it equaled m_1, and then calculating the total time lived in that year.

Table 6.1. Life Table for Male Dall Mountain Sheep (adapted from Murie 1944). The symbols at the head of each column are explained in the text.

a years	L_a	D_a	m_a	$\overline{L}_a$	$\bar{e}_a$ years
0	1000	450	.450	680	4.7
1	550	97	.176	502	7.3
2	453	13	.028	446	7.7
3	440	12	.027	434	6.9
4	428	12	.028	422	6.1
5	416	14	.034	409	5.3
6	402	15	.037	394	4.5
7	387	30	.078	372	3.6
8	357	45	.126	334	2.9
9	312	57	.183	284	2.2
10	255	56	.220	227	1.6
11	199	123	.618	138	0.9
12	76	71	.934	40	0.6
13	5	5	1.000	2	0.4
14	0				
TOTALS	5,280	1,000		4,684	

The expectation of further life, the $\bar{e}_a$ column, is derived by calculating the average number of years lived after each exact age. In symbols,

$$\bar{e}_a = \frac{1}{L_a} \Sigma_{j=a}^{\omega} \overline{L}_j, \tag{6.15}$$

where ω is the oldest age. The initial value, $\bar{e}_0$, is the average longevity or length of life of the cohort.

Life tables were originally developed as a means of arriving at the last column, the expectation of further life at each age, because of the necessity of this information for the life insurance business. The expectations are rarely used, however, in ecological studies, and except for the inherent interest lying in these figures, they are hardly worth the labor of calculating them. Since the main purpose of the $\bar{L}_a$ column is for obtaining the $\bar{e}_a$ figures, only the first four columns are of much significance to ecologists. These can be reduced to three to make a form of life table that will prove useful in later chapters and that is illustrated in table 6.2.

Table 6.2. Survivorship Table (abbreviated) for Male Dall Mountain Sheep. The symbols at the head of each column are explained in the text. The data are in table 6.1.

a years	s_a	l_a
0	.550	1.000
1	.824	.550
2	.972	.453
3	.973	.440
.	.	.
.	.	.
.	.	.
13	.000	.005
14	.	.000

The column headings are:
a : age at beginning of interval, as before.
s_a: age-specific survival rate, $= L_{a+1}/L_a$.
l_a : probability of a newborn attaining age a, $= L_a/L_0$.

6.4 THE LIFE TABLE IN A STATIONARY POPULATION

A stationary population is one with unchanging and equal birth and death rates and it is also closed to immigration and emigration, so it is in a steady state. These conditions may appear to be unrealistic, but since populations never grow or decline indefinitely, there are likely to be times when a population approaches the stationary condition. The concept of a stationary population is useful in understanding the relationships between some population characteristics and the life table.

The life table of the mountain sheep presented in the preceding section will be used as an illustration and will be assumed to apply to both sexes. Deaths occur throughout the year, but the young are born in a short lambing season, and we will look at the population at this time of the year. The situation where the young are born throughout the year will be considered later.

Imagine a stationary population of these sheep of such a size that 1000 are born each lambing season to replace 1000 of the population that died during the year just preceding. A year later the 1000 born will have dwindled to 550, when another 1000 lambs will be produced. Two years later the first group will be down to 453, the second will have become 550, and 1000 new lambs will be present. Continuing this procedure step by step one can easily see that the age composition of the population at the close of the lambing season will be the same as the L_a column of the life table. The size of the population then is ΣL_a, or 5,280.

The numbers dying of each age are in the D_a column. The age-specific mortality rates are, of course, in the m_a column, and the age-specific survival rates can be calculated from $s_a = 1 - m_a$ or from L_{a+1}/L_a.

The annual general population rates and other characteristics may be found as follows (illustrating with figures from table 6.1). The general mortality rate, $m = \Sigma D_a/\Sigma L_a = 1000/5280 = 1894/10{,}000$. Since L_0 equals the sum of the D_a, $m = L_0/\Sigma L_a$, which is the proportion of newborn in the population. The general survival rate is most easily calculated from $(1 - m) = 8106/10{,}000$. The general death rate, $d = \Sigma D_a/\Sigma \overline{L}_a = 1000/4684 = 2135/10{,}000$. The general birth rate in this case is best defined as the number born to the number alive just before the reproductive season: $b = L_0/(\Sigma L_a - L_0) = 1000/4280 = 2336/10{,}000$. Although this is a stationary population, b and d are only approximately equal because they are measured with respect to different population bases.

The average longevity, since the sum of the D_a equals L_0, is the reciprocal of the death rate; $\bar{e}_0 = \Sigma \overline{L}_a/L_0 = 1/d = 4.7$ years.

Turning now to the situation in which both births and deaths occur throughout the year, as in man, we will find some characteristics of such a stationary population are in different parts of a life table than in the case illustrated by the mountain sheep. At any one time in the year, the number of individuals between ages a and $(a + 1)$ will be best estimated by $\overline{L}_a$, the average number alive during the interval. Therefore the age distribution will equal the distribution of the $\overline{L}_a$, and the size of the population will equal $\Sigma \overline{L}_a$.

The general mortality rate will be, when births occur throughout the year, $m = \Sigma D_a/\Sigma \overline{L}_a$. This is the same as the general death rate, and since $\Sigma D_a = L_0$, it equals the birth rate, $m = db$.

The general survival rate, equaling $(1 - m)$, is also measured with respect to the sum of $\bar{L}_a$. The age-specific mortality and survival rates, however, will be measured as in the previous case because they refer to individuals of a certain age; for example, $s_a = L_{a+1}/L_a$ as before.

For purposes of summary and reference, the equations that relate the life table and a stationary population are set forth in table 6.3.

Table 6.3. Population Characteristics in the Life Table of a Stationary Population.

Characteristic (all rates are annual)	A	B
Proportion in each age class, C_a	$\frac{L_a}{\Sigma L_a}$	$\frac{\bar{L}_a}{\Sigma \bar{L}_a}$
Age-specific survival rate, s_a	$\frac{L_{a+1}}{L_a} = \frac{C_{a+1}}{C_a}$	$\frac{L_{a+1}}{L_a} = \frac{C_{a+1}}{C_a}$
General mortality rate, m	$\frac{\Sigma D_a}{\Sigma L_a} = \frac{L_0}{\Sigma L_a}$	$\frac{\Sigma D_a}{\Sigma \bar{L}_a} = \frac{L_0}{\Sigma \bar{L}_a}$
General death rate, d	$\frac{\Sigma D_a}{\Sigma \bar{L}_a} = \frac{L_0}{\Sigma \bar{L}_a}$	$\frac{\Sigma D_a}{\Sigma \bar{L}_a} = \frac{L_0}{\Sigma \bar{L}_a}$
General reproductive rate, b	$\frac{L_0}{\Sigma L_a - L_0}$	$\frac{L_0}{\Sigma \bar{L}_a} = \frac{\Sigma D_a}{\Sigma \bar{L}_a}$
Average longevity, $\bar{e}_0$	$\frac{\Sigma \bar{L}_a}{L_0} = \frac{\Sigma \bar{L}_a}{\Sigma D_a}$	$\frac{\Sigma \bar{L}_a}{L_0} = \frac{\Sigma \bar{L}_a}{\Sigma D_a}$
Equalities	$m = b/(b + 1)$	$m = d = b$
	$m = C_0$	
	$s = 1 - m$	$s = 1 - m$
	$\bar{e}_0 = \frac{1}{d}$	$\bar{e}_0 = \frac{1}{m} = \frac{1}{d}$

Column A: Stationary population with births occurring at one time of the year, characteristics determined at that time.
Column B: Stationary population with births occurring throughout the year.

6.5 TYPES OF SURVIVORSHIP

A graphic record of the history of a cohort is made by plotting the number of survivors of each age (L_a) against age (a) or time, forming what is called a survivorship curve. There are advantages to using a logarithmic scale on the vertical axis for the number of survivors and an arithmetic scale on the horizontal axis for age; if the survival rate is constant, L_a will equal $L_0 s^a$, and $\log L_a = \log L_0 + a \log s$, so the line will be straight and nearly horizontal if the survival rate is very high or declining steeply if the rate is low. If not straight, a curve bending downward indicates a declining survival rate, and one bending upward an improving rate.

Three idealized types of survivorship are illustrated in figure 6.3. The simplest, termed ecological survivorship, is a straight line because

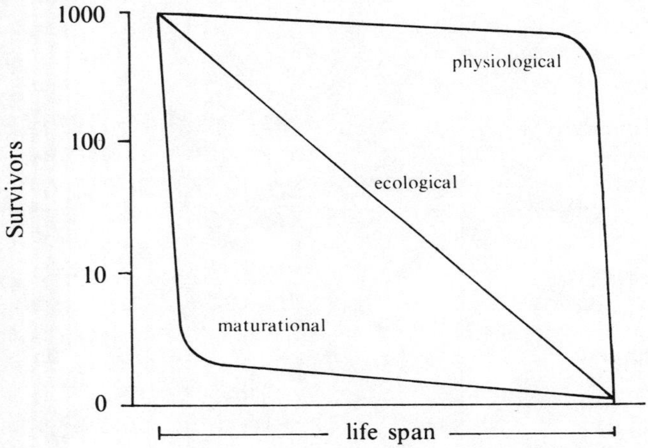

Figure 6.3. Types of survivorship curves. The scale for the number of survivors is logarithmic.

the survival rate is constant. This is called "ecological" because survival is independent of the age of the animals, and deaths are therefore due to environmental causes such as accident and predation. Since the probability of death is independent of age, the expected longevity from any age will be the same; this however would not be biologically realistic for very old individuals.

Physiological survivorship is characterized by high survival during most of the cohort's history, until old age is reached when all the survivors die almost simultaneously. The term "physiological" is used because death is almost entirely due to aging. The average longevity approaches the life span of the cohort. A species like wheat, where all the plants in a field ripen and die almost simultaneously, appears to possess this type of survivorship, and it seems to be one of the objectives of mankind.

The third simple type of survivorship shows a very poor survival soon after birth, which later improves to a high survival at the attainment of adulthood; this is therefore called "maturational" survivorship. The average longevity of species of this type is a very small fraction of the life span. Oysters and many endoparasites probably possess this type of survivorship, although there is very little data to confirm this supposition.

Of course, not every species should be expected to possess one or another of the three simple types of survivorship illustrated in figure 6.3. In fact, all of those which have been sufficiently studied show survivorship curves illustrating a combination of types. Figure 6.4 shows three examples of observed survivorship; on each is plotted the history of a typical cohort over the life span with one line showing the survivorship and a second showing the age-specific mortality rates.

The survivorship of birds has been studied more than any other group, and the subject has been reviewed by Farner (1955) and Hickey (1952). Mortality rates are high for the egg, nestling, fledgling, and early immature stages, and then decrease as adulthood is attained and become independent of age (see fig. 6.4A). Their survivorship is a combination of maturational and ecological. Long-lived species, such as some sea birds, apparently have an increase in mortality rates at very old ages (Botkin and Miller 1974), which would deflect the survivorship curve downward near the end of the life span.

Caughley (1966) and Spinage (1972) studied the mortality patterns of mammals and found that the survivorship for several species of large mammals is similar to that of the mountain sheep illustrated in figure 6.4B, which was constructed from the data in table 6.1. There is high mortality in the very young, followed by a pattern which approaches a physiological survivorship; that is, low mortality in the young adults and very high in the old. This is also the general pattern possessed by modern man.

A composite survival curve for several species of insects with complete metamorphosis is shown in figure 6.4C (data from Campbell 1969; Harcourt 1969 and 1971; McKnight 1971; Morris and Miller 1954; and Mukerji 1967). It indicates low mortality during the inactive egg and pupal stages and high mortality when individuals are moving and dispersing.

About 120 citations to papers on mortality, mostly studies on species of vertebrates, are included in a bibliography on quantitative ecology (Schultz *et al.*, 1976, pp. 299–308).

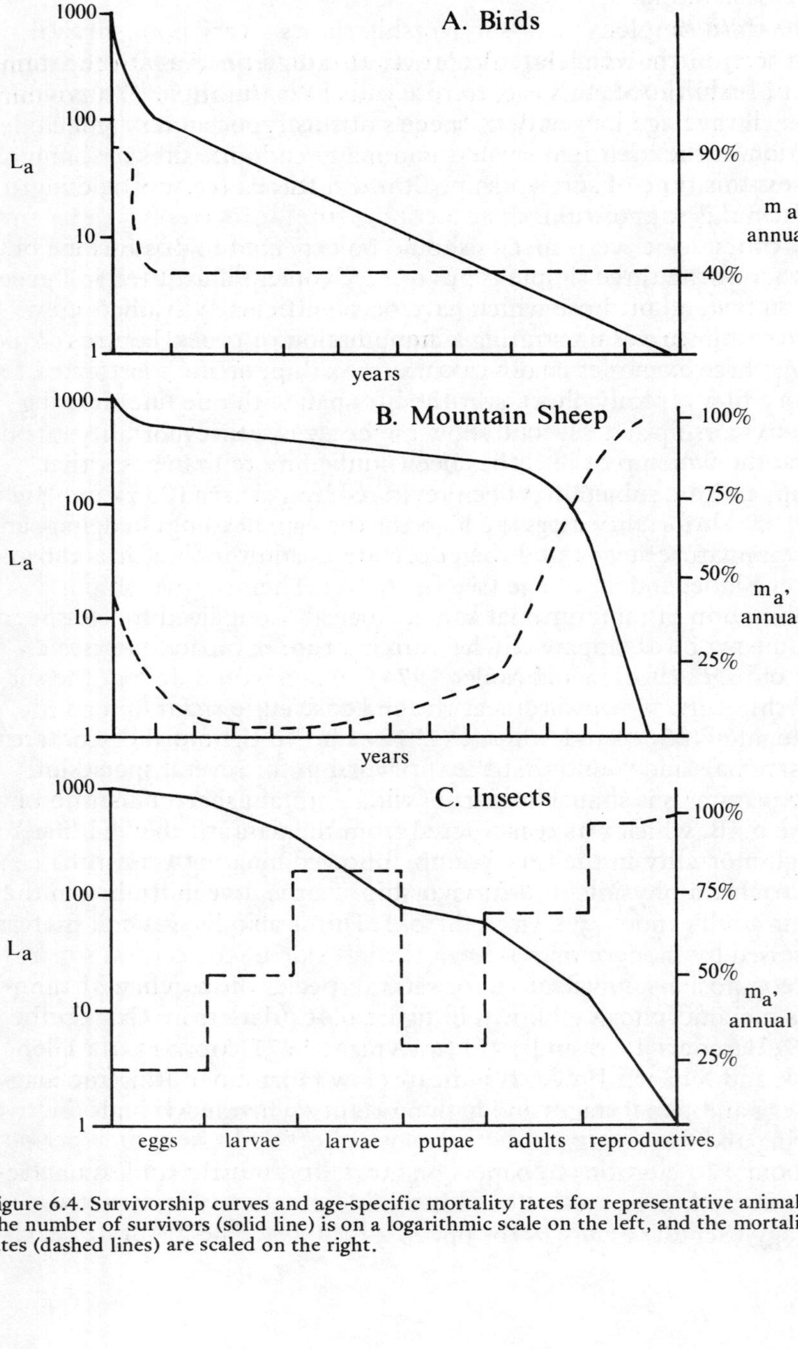

Figure 6.4. Survivorship curves and age-specific mortality rates for representative animals. The number of survivors (solid line) is on a logarithmic scale on the left, and the mortality rates (dashed lines) are scaled on the right.

6.6 CALCULATIONS OF SPECIFIC SURVIVAL AND MORTALITY RATES

Introduction

This section concerns the calculation of rates from data on the number of survivors of each age, or of deaths at each age, or of a combination of the two kinds of data. Some statistical procedures are included. The subject is organized around the pattern of a life table to simplify the explanations. There are two general methods for constructing a life table. One involves tracing a cohort from its beginning to its end, thus obtaining a series of L_a or of D_a values. A life table formed in this way is termed a *generation life table* (Spiegelman 1968). The data for the second method come from the number of individuals attaining each age and the number of these dying in a period such as a calendar year. From these can be calculated an age-specific mortality rate. Such a life table is called a *current life table* (Spiegelman 1968) because the m_a values present the age-specific mortality for the period when the data were collected. Most human life tables are current because a generation life table would require too many years to prepare, and because demographers and life insurance companies are interested in today's rates and expectation of life rather than those of years ago.

Synonyms for the types of life table are, for generation: age-specific, dynamic, and horizontal; for current: time-specific, static, and vertical (Deevey 1947, Hickey 1952).

Some of the problems arising in the construction of a human life table, or of any species where births occur throughout the year, are illustrated in the "Lexis diagram" of figure 6.5. Each diagonal line represents an individual beginning with birth at age 0, at the top of the diagram. Its age increases linearly with time until the "lifeline" terminates with death. Consider the lines terminating within the central square of this figure, whose deaths occur between times 1 and 2 and between exact ages 1 and 2. Some of these individuals passed age 1 in the previous period (crossing the left side of the central square) while others did so within this period (crossing the top line of the square). The demographers' solution to this problem—to obtain the "best" age-specific mortality rate—is rather involved (Keyfitz 1968, Spiegelman 1968). Since most wild species of animals breed seasonally, the situation can be simplified by considering each reproductive season as the beginning of both a new cohort and a new time period. If the cohorts live through more than one reproductive season and thus overlap, the situation is pictured in figure 6.6. The top row, the L_{c0}, represents the original size of each cohort. The history of each

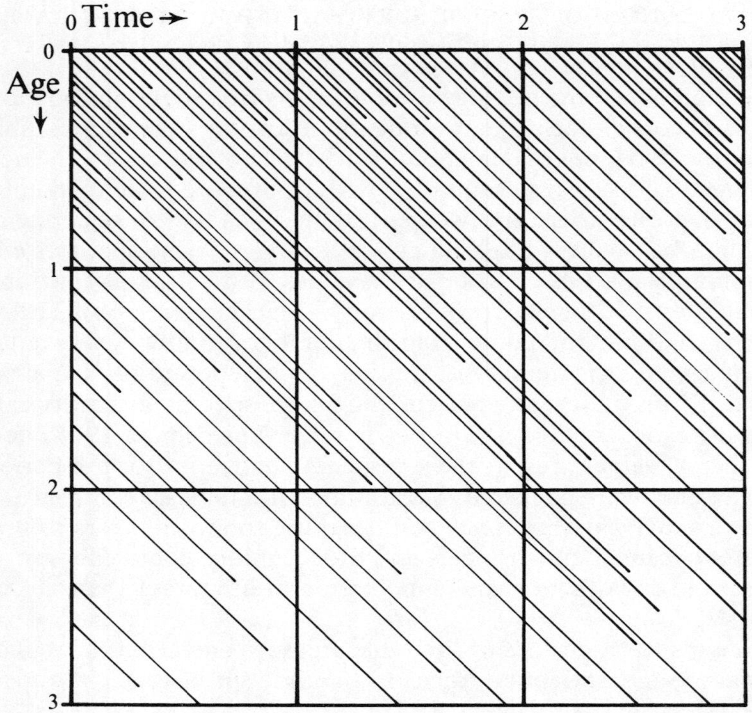

Figure 6.5. A Lexis diagram. Each diagonal line represents the increase in age of one individual with time from birth to death.

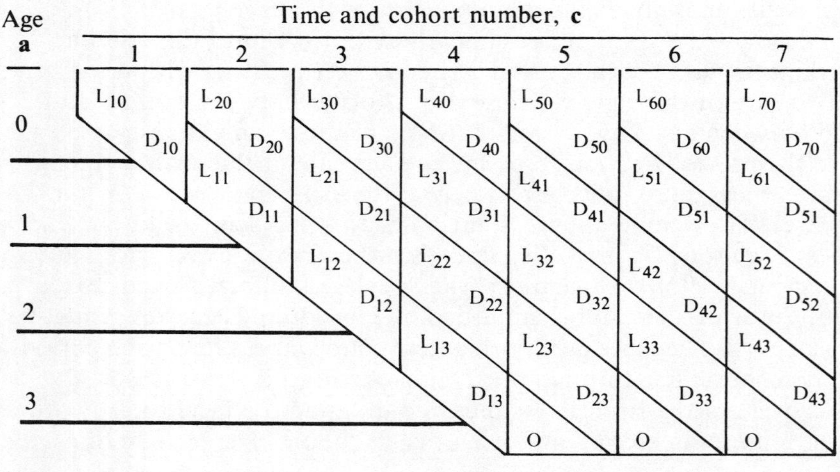

Figure 6.6. History of cohorts for a species with a short reproductive season. Time periods begin with the production of a new cohort.

cohort is shown by the diagonal line of L_{ca}, the number of that cohort attaining age a, and of D_{ca}, the number of that cohort dying between ages a and $(a + 1)$. A vertical column of L_{ca} represents the age composition of the living at the beginning of a particular time period, and a vertical column D_{ca} represents the age composition of those dying within that time period.

The notation used in this chapter for indicating the age of an individual differs from that used by some other authors. Here the symbol D_a, for example, refers to the number dying between exact ages a and $(a + 1)$, whose most recent birthday was at age a; the initial value of a is therefore 0. Others (e.g., Seber 1972) use D_a for the number dying in their "ath year of life," not having attained age a, so the first a is 1. This difference in notation must be understood when formulas used by different authors are compared.

Methods used in determining the age of individual animals are referred to in chapter 4, section 4.2. The use of marked individuals in estimating age-specific rates is discussed at the conclusion of this section.

The formulas given in the following subsections are not derived or proved because many are obvious from their definitions. Anyone wishing to prove the not-so-obvious formulas will find the following equations to be useful. They are based on l_a, the probability of surviving to age a; the index c refers to a particular cohort:

$$l_a s_a = l_{a+1};$$

$$L_{ca} = L_{c0} l_a;$$

$$D_{ca} = L_{c0} l_a m_a;$$

$$\Sigma_{a=0}^{\omega} l_a m_a = 1.$$

Survivors of One or More Cohorts

A complete record of the survivors of a cohort provides a sequence of the L_a, and the age-specific survival or mortality rates may be calculated directly from these data as described in a preceding section on the life table. Most data on natural populations are, however, only a sample of the survivors, and if the probability of an individual being included in the sample depends on the individual's age, this probability must be considered. Let the probability of a survivor of age a being sampled (caught or counted) be symbolized by $P(L_a)$, then the expected sample size of age a is $L'_a = L_a P(L_a)$. To estimate the survival rate from the ratio of the sample numbers will require a sampling correction, as follows. For

$$s_a = \frac{L_{a+1}}{L_a},$$

$$\hat{s}_a = \frac{L'_{a+1}}{L'_a} \frac{P(L_a)}{P(L_{a+1})} = \frac{L'_{a+1}}{L'_a} (sci), \tag{6.16}$$

where $(sci) = P(L_a)/P(L_{a+1})$ is the sampling correction for this ratio.

If the data are a succession of samples of the survivors, a series of L'_a, there are advantages to adjusting this series to a base equal to the L'_0, the sample of the youngest and largest age group. The sampling correction for this purpose is

$$(sca) = \frac{P(L_0)}{P(L_a)}, \tag{6.17}$$

and the adjusted number of survivors is $L''_a = L'_a(sca) = L_a P(L_0)$. Thus the data are transformed to a series of numbers: L''_0 $(=L'_0)$, L''_1, L''_2, . . . L''_ω, which are proportional by a factor of $P(L_0)$ to the actual number of survivors and which may be used as a survivorship column. (Here and in the following pages L_a symbolizes the actual number of survivors of age a; L'_a, the observed sample of age a; and L''_a, the latter number adjusted by sampling and cohort size corrections. Similar notation will be used with D_a, the dead of each age.)

A sampling correction is probably easier to estimate than a sampling probability because the former, being a ratio, needs only the relative sizes of the two probabilities. In many cases it is difficult to obtain much information on the L_0, the newborn, so in these a sampling correction must be based on another age group, the youngest for which the relative probability of being sampled can be estimated. Let v stand for this age; then the sampling correction is

$$(sca) = \frac{P(L_v)}{P(L_a)}. \tag{6.18}$$

If the sampling is believed to be independent of age (i.e., if the assumption of equal catchability holds), the correction equals 1. In many sampling methods, and with the exception of very young animals, this assumption is valid and the sampling correction need not be used. It will be included, however, as (sci) or (sca) in most of the formulas that follow to keep them as general as possible.

Numbers of survivors of the same age from different cohorts may be combined to yield larger samples. Sampling corrections may be applied either before or after the summing. The sums can be used to calculate rates, however, only if the sums come from identical cohorts. Using figure 6.6, for example, s_0 can be calculated from the sums of

L_0 and L_1 across cohorts 1 through 6, and s_2 can be calculated from the sums L_2 and L_3 across cohorts 1 through 4 only. (Data from additional cohorts may be included if the sums are adjusted as explained in connection with equation [6.42].)

Ex. 6.5. Age-specific survival rates estimated by combining data from different cohorts. A record of survivors from four cohorts is used to calculate age-specific survival rates. The L_0, the newborn, are unknown, so the survivors of age 1 are used as the base of each cohort. The data are tabulated as in figure 6.5.

Age	Time or Cohort Number 1	2	3	4	5
0	?	?	?	?	?
1		37	65	40	48
2			31 (22)	46 (32)	34 (24)
3				17 (12)	36 (25)
4					10 (7)

NOTE: Studies indicate that the L_1 are less likely to be sampled than the older individuals, and that a sampling correction (eq. 6.18) of 0.7 should be used for ages 2, 3, and 4. The adjusted data are indicated in parenthesis in the above table.

Combining cohorts, the age-specific survival rates based on the adjusted data are:

$$\hat{s}_1 = \frac{22 + 32 + 24}{37 + 65 + 40} = .55;$$

$$\hat{s}_2 = \frac{12 + 25}{22 + 32} = .68;$$

$$\hat{s}_3 = \frac{7}{12} = .58.$$

A graph of the survivorship of a cohort is expected to be a smooth curve depicting gradual changes in the rates, but the plot of a survivorship estimated from samples is almost certain to show irregularities. Actuaries have developed several procedures for smoothing or "graduating" data from a cohort; descriptions of these may be found in Keyfitz (1968) or Spiegelman (1968). Methods of estimating a survival rate independent of age, or constant over several ages, are described in a later subsection.

Estimation of Survival Rates from Cohorts when Data are Missing

If the numbers from one or more age classes are unknown, a survival rate may be calculated over the ages from which data are missing by assuming a constant rate for the interval. If s is the probability of surviving through one period and if the rate is constant through k such periods, then

$$(\text{survival over } k \text{ periods}) = s^k,$$

or

$$s = \sqrt[k]{(\text{survival over } k \text{ periods})}. \qquad (6.19)$$

For example, if the survivorship data are missing for age $(a + 1)$, and if it can be assumed that the survival rate is constant over two periods, then

$$s_a = s_{a+1} = \sqrt[2]{L_{a+2}/L_a}.$$

Larger gaps may be treated in the same way using higher roots, but the assumption of equal age-specific rates may be less tenable.

Ex. 6.6. Estimation of survival rate when a year's data are missing. If in the record of example 6.5 the data on the L_2 were missing, an estimate could be made by assuming that $s_1 = s_2$:

$$\hat{s}. = \sqrt{\frac{12 + 25}{37 + 65}} = .60.$$

Problems of similar nature arise when observations are made over only segments of a stage of the life cycle, and an estimate is desired of survival over the whole stage. For example, the "nest hatching success" of birds is the survival rate of nests over the incubation period. Data for estimating this are usually fragmentary because nests are often not found until incubation has started, and then they are not examined daily by the observer. This particular problem is analyzed by Mayfield (1975). The solution may be described in general terms as follows. Let s be the survival rate per individual-period (e.g., per nest-day), and let $m = 1 - s$ be the death or failure per individual-period. As discussed in the previous chapter, a rate may be thought of as the number of events observed divided by the total time lived during which those events could have been observed. Therefore m, as just defined, is:

$$m = \frac{\text{number of deaths or losses observed}}{\text{total individual-periods in which these could have occurred}}.$$

The denominator of this ratio is the total over all individuals (nests or whatever) of the number of periods (e.g., days) in which each individual was known to be alive, and therefore in which it could have died. A problem naturally arises when a death or loss occurs at some time during a gap in the observations; the best approximation is to assume that it occurred at the midpoint of the gap as this will introduce very little error if the probability of survival over each period is high.

Once *m* is calculated as above from the data, 1 - *m* gives the survival rate over each period, and from the number of such periods in the life history stage the overall survival can be calculated from equation (6.19).

Ex. 6.7. Estimation of a survival rate from partial data. Ten nests were observed over varying periods, none being observed over the complete incubation period of fourteen days.

Nest Number	Inclusive Dates Known to Be Surviving		Date of Loss x = no loss observed	Number of Nest-days Surviving
1	2 - 8		x	7
2	2 - 9		9 or 10	8.5
3	2 - 14		x	13
4	5 - 8		8	4
5	5 - 8	Nest empty on 12:	10, assumed	6
6	5 - 9	Nest empty on 12:	10.5, assumed	6.5
7	6 - 18		x	13
8	6 - 12		x	7
9	6 - 8		x	3
10	6 - 8	Nest empty on 12:	10, assumed	5
		Number of losses:	5	Total: 73

$$m = \frac{5}{73} = .0685$$

$$s = 1 - .0685 = .9315,$$

survival over 1 day.

$$s^{14} = .37, \qquad \text{(eq. 6.19)}$$

survival over the fourteen-day incubation period, or nest hatching success.

Deaths from One or More Cohorts

A complete record of the deaths at each age from a cohort with no survivors can be used as described in the section on the life table to reconstruct the cohort's history. The sum of all the D_a equals the cohort's L_0. In figure 6.6, cohorts 1 through 4 could be treated in this way, but not the later cohorts which are not complete by year 7. If samples of the deaths at each age are the data, the probability of a dead individual being sampled or counted must be considered. The argument is the same as in the previous subsection, and the symbols are parallel. The sampling correction to adjust a series of D'_a to a base of D'_0 is

$$(sca) = \frac{P(D_0)}{P(D_a)}, \qquad (6.20)$$

and the adjusted number dead at each age is $D''_a = D'_a\,(sca) = D_a P(D_0)$. The adjusted values are proportional to the true numbers. These can be used to reconstruct a history of the cohort because

$$\Sigma^{\omega}_{a=0}\, D''_a = P(D_0)\,\Sigma^{\omega}_{a=0}\, D_a = P(D_0)L_0.$$

The constant $P(D_0)$ will not affect the proportions and the usual life table methods can be used.

The earlier remarks about the difficulty of estimating sampling probabilities for the youngest age classes also apply here. The same modification can be used, that is, to start the cohort history with the youngest age class for which sampling probabilities or corrections can be estimated. If this age is designated v, this symbol is substituted for 0 in equation (6.20). This same procedure can be followed if the numbers of deaths are unknown in the age classes younger than age v.

Ex. 6.8. Deaths from one cohort used as the basis of a life table. Assumed data are: year 1956, 59 juveniles known dead; 1957, 44 one-year known dead; 1958, 94 two-year known dead; 1959, 40 three-year known dead; none older reported. The probability of finding dead juveniles is believed to be half that of finding older ages; from equation (6.20), $(sca) = .5$, for $a = 1$, 2, and 3.
The adjusted data, D''_a, and the calculated L_a and m_a are:

a	D''_a	L_a	m_a
0	59	148	.40
1	22	89	.25
2	47	67	.70
3	20	20	1.00
4	0	0	
	148		

The preceding paragraphs are concerned with analyzing the death records from complete or "extinct" cohorts, such as the first three cohorts of figure 6.6. Deaths from incomplete cohorts, those in which survivors are still present, do not have enough information to allow estimating an L_0 and a survivorship column. Another method will work under one condition and is shown as follows. The ratio of deaths of two successive ages is $D_{a+1}/D_a = m_{a+1}L_{a+1}/m_aL_a$. If the two age-specific mortality rates, m_a and m_{a+1}, are equal,

$$s_a = \frac{L_{a+1}}{L_a} = \frac{D_{a+1}}{D_a} = \frac{D'_{a+1}}{D'_a} \text{ (sci).} \tag{6.21}$$

This method has some weaknesses, which are discussed by Cormack (1970) and Seber (1972), but it may frequently be useful to obtain an approximation of the survival rate from samples of deaths at successive ages.

The number of deaths at each age from different cohorts can be combined to make larger samples. The procedures and conditions are the same as those for combining numbers of survivors, described in the previous subsection.

An age-specific mortality rate may be estimated from the number of survivors at age *a* and the number of deaths between ages *a* and (*a* + 1), or from a sample of these if a sampling correction is known:

$$m_a = \frac{D_a}{L_a} = \frac{D'_a P(L_a)}{L'_a P(D_a)} \; . \tag{6.22}$$

Since the probabilities of sampling living individuals are likely to be very different from those of sampling dead, the correction cannot be assumed to equal 1. The age-specific mortality rate may also be calculated from the number of deaths in the cohort of age *a* and older. Assuming in this case that the sampling correction is 1,

$$m_a = \frac{D_a}{\sum_{i=a}^{\omega} D_i} \; . \tag{6.23}$$

Rates from Age Composition of the Living

The age composition of the living at any one time is illustrated in figure 6.6 by a vertical column of L_{ca} for year 5, 6, or 7. If the population is stationary, and if all the living have been counted or if each age has equal probability of being sampled, the L_a column can be treated as if it came from one cohort. The relations between the structure of a stationary population and the survivorship column of a life table are discussed in section 6.4 above. If the population size was changing, however, corrections must be made for this fact because

Ex. 6.9. Deaths from a cohort used directly to calculate age-specific mortality rates. The adjusted data in the table of example (6.8) can be used to illustrate equation (6.23):

$$m_0 = \frac{59}{59 + 22 + 47 + 20} = .40;$$

$$m_1 = \frac{22}{22 + 47 + 20} = .25;$$

$$m_2 = \frac{47}{47 + 20} = .70;$$

$$m_3 = \frac{20}{20} = 1.00 \, .$$

each age group in a vertical column belongs to a different cohort, and the cohorts' original sizes were unequal. A good way to correct data for variations in cohort size is to adjust the figures by using the L_{c0} of that time, the youngest age group in the population, as the base for the adjustment. The result will be, for example, that the corrected number of living at time 5 in figure 6.6 will be equal to the survivors from a cohort of initial size equal to L_{50}. There are two ways of doing this. The first requires that the relative sizes of the original strengths of the cohorts be known. The cohort correction for each age a is

$$\text{cohort correction, (cc)} = \frac{L_{c0}}{L_{(c-a)0}} \, . \qquad (6.24)$$

For example, using figure 6.6, the cohort correction for $a = 2$ in year 5 is L_{50}/L_{30}. The second method (basically the same) is useful if it is known that the population has been growing or diminishing at a steady rate. In this case the ratio between the number of newborn at any time to the same age in the previous year is a constant: $L_{c0}/L_{(c-1)0} = g$, and the cohort correction for each age a is

$$\text{(cc)} = g^a \, . \qquad (6.25)$$

A sampling correction may also be needed. Since L_{c0} is the base for the cohort correction, the sampling correction is the (sca) of equation (6.17). Multiplying the observed number of age a, $L'_{(c-a)a}$, by both the cohort correction and the sampling correction produces the corrected number of age a: $L'_{(c-a)a}(cc)(sca) = L''_a = \hat{l}_a L'_{c0}$. The symbol $\hat{l}_a$ is the estimated survivorship to age a. Expanding this equation gives:

Data	Corrected Number	Equivalent
L'_{c0}	L''_0	L'_{c0}
$L'_{(c-1)1}$	L''_1	$\hat{l}_1 L'_{c0}$
$L'_{(c-2)2}$	L''_2	$\hat{l}_2 L'_{c0}$
-	-	-
$L'_{(c-a)a}$	L''_a	$\hat{l}_a L'_{c0}$

The corrected age composition can be used to calculate survival or mortality rates just as if the figures were the survivorship column of a life table.

Many students who have estimated rates from samples of the ages of the living have believed or assumed that the population was stationary so that the cohort correction was 1. They have also begun with the youngest age class at which the sampling probability did not change with age so that the sampling correction was also 1. Under both assumptions the age composition can be treated as representing a cohort beginning with the youngest age to which the assumption of equal catchability applies.

The numbers of survivors of the same age, from different cohorts and different years, may be combined to give larger samples. Denoting the sum across the same range of years by the symbol

$$\Sigma_c,$$

the cohort correction is

$$(cc) = \frac{\Sigma_c L_{c0}}{\Sigma_c L_{(c-a)0}} . \tag{6.26}$$

The corrected number of survivors is then

$$L''_a = (\Sigma_c L'_{(c-a)a})(cc)(sca) = \hat{l}_a \Sigma_c L'_{c0} . \tag{6.27}$$

The procedure is to sum the survivors of one age across two or more years (cohorts); sum the L_0 from which these survivors came; sum the L_0 for the years in which these survivors reached age a; determine the sampling correction if any; then apply equations (6.26) and (6.27) to the calculation of L''_a. For an example from figure 6.6, for $x = 3$ and the years 5, 6, 7 (with the assumption that the sampling correction is 1):

$$L''_3 = \Sigma_c L'_{(c-3)3} \frac{\Sigma_c L_{c0}}{\Sigma_c L_{(c-3)0}}, \; c = 5, 6, 7$$

$$= (L_{23} + L_{33} + L_{43}) \frac{L_{50} + L_{60} + L_{70}}{L_{20} + L_{30} + L_{40}}.$$

The corrected sums can be used in the same way as numbers from an age composition at one time. An alternative cohort correction can be used if the population has been growing or diminishing at a steady rate; in such a case equation (6.26) can be replaced with $(cc) = g^a$ (see eq. [6.25]).

Ex. 6.10. Age composition of the living used to form a corrected L_a column. The numbers of each age counted in two successive years were:

a	1	2
0	72	81
1	69	74
2	48	54
3	28	31
4	7	6

The probability of counting animals in the youngest age class is estimated to be .67 of that of older animals, so the sampling correction $(sca) = .67$ (eq. [6.40]) for $a = 1, 2, 3$, and 4. The population has been increasing at a steady rate of 10 percent per year, or $g = 1.1$, making a cohort correction (eq. [6.25]) of $(cc) = 1.1^a$. Using equation (6.27):

$L_0 = (72 + 81) \times 1 \times 1 = 153;$

$L_1 = (69 + 74) \times 1.1 \times .67 = 105.4;$

$L_2 = (48 + 54) \times 1.1^2 \times .67 = 82.7;$

$L_3 = (28 + 31) \times 1.1^3 \times .67 = 52.6;$

$L_4 = (7 + 6) \times 1.1^4 \times .67 = 12.8.$

Rates from Age Composition of the Dead

The number of animals that died in a calendar year or similar period and that were between ages a and $(a + 1)$ is counted in the D_a, and the proportions of the different ages make up the age composition of the dead. A vertical column of D_{ca} in figure 6.6 symbolizes this.

Before the methods for analyzing such data are described, an important point needs to be made. If the animals have been killed by a cause acting independently of the animals' age, then the number of dead at each age is really a sample of the age composition of the living. Catch curves in fisheries are examples (Ricker 1975) because the fishing gear catches individuals above a certain size with equal probability. Kill curves (Eberhardt 1971) are similar sets of data acquired from hunters' kill of game, which within certain limits are independent of the animals' age. Other examples might be animals killed by a sudden catastrophe, a poisoning campaign, or by automobiles. If all deaths are independent of age, survivorship is ecological (see earlier section on types of survivorship), and the age composition of the living and of the dead will be proportional to each other. But if this is not the case, and samples of dead are obtained by some method independent of age, these should be treated as being a sample of the age composition of the living and should be analyzed by the methods of the preceding subsection. This point and other sources of error and bias in estimating survival rates from the age composition of the dead, especially of mammals, are discussed further by Caughley (1966).

Data from the age composition of the dead are adjusted by methods identical to those used for that of the living. The cohort correction may be

$$(cc) = \frac{D_{c0}}{D_{(c-a)0}}, \tag{6.28}$$

or since $D_{c0} = m_0 L_{c0}$, equation (6.24) or (6.25) may be used. The sampling correction is equation (6.20). Multiplying the sample number by the two corrections (using eq. [6.24] for the cohort correction) yields the corrected number

$$D''_a = D'_{(c-a)a}(cc)(sca) = \widehat{l_a m_a} L_{c0} P(D_0). \tag{6.29}$$

The $\widehat{l_a m_a}$ is the probability of surviving to age a and dying before age $(a + 1)$, estimated from the numbers of the $(c - a)$ cohort. An expansion of equation (6.29) gives

Data	Corrected Number	Equivalent
D'_{c0}	D''_0	D'_{c0}
$D'_{(c-1)1}$	D''_1	$\widehat{l_1 m_1} L_{c0} P(D_0)$
$D'_{(c-2)2}$	D''_2	$\widehat{l_2 m_2} L_{c0} P(D_0)$
-	-	-
$D'_{(c-\omega)}$	D''_ω	$\widehat{l_\omega m_\omega} L_{c0} P(D_0)$

The sum over all corrected numbers, from 0 to ω,

$$\Sigma_{a=0}^{\omega} D''_a = L_{c0}\, P(D_0).$$

Therefore a survivorship column can be constructed from the D''_a if all ages are included (e.g., for the years 4, 5, 6, and 7 of figure 6.6).

Ex. 6.11. Samples of dead at each age used to construct a corrected L_a column. In this hypothetical example individuals found dead in each of two calendar years were classified according to their ages (a = 0 for those less than 1 year). None four years old or older were found. The number are:

a	1974	1975
0	28	31
1	4	6
2	4	1
3	3	4

The probability of sampling each age is believed equal, so no sampling correction is necessary. A cohort correction, however, is needed, because it is known that in 1973 only half the usual number of young were added to the population, so $(cc) = 2$ when $(c - a) = 1973$; otherwise $(cc) = 1$. Using equation (6.29) while omitting the sampling correction:

1974: $D''_0 = 28 \times 1 = 28$	1975: $D''_0 = 31 \times 1 = 31$
$D''_1 = 4 \times 2 = 8$	$D''_1 = 6 \times 1 = 6$
$D''_2 = 4 \times 1 = 4$	$D''_2 = 1 \times 2 = 2$
$D''_3 = 3 \times 1 = 3$	$D''_3 = 4 \times 1 = 4$

Adding the corrected numbers to obtain a larger sample, a survivorship column is calculated.

a	D_a	L_a
0	59	86
1	14	27
2	6	13
3	7	7
	86	

If the numbers of deaths in the youngest age classes are missing, and the record is complete only for ages v to ω, the 0 in equation (6.29) can be replaced with v and the survivorship constructed for the later years of the cohort. Another alternative is to calculate the age-specific mortality for each age v and older using the corrected D''_a with equation (6.23).

If the older age classes are not included in the samples of the dead, the above methods cannot be used. Any two successive numbers can be used to estimate a survival rate, if that can be assumed to be equal for the two years, by using the corrected numbers in equation (6.21), which has been discussed and qualified in an earlier paragraph.

Numbers of dead of the same age from different cohorts may be added and adjusted by cohort and sample corrections in steps parallel to those described for combining numbers from the age composition of the living (eq. [6.27]). For example, from figure 6.6, with $a = 3$ and the years 4, 5, and 6, and with the sampling correction equaling 1:

$$D''_3 = \Sigma_c D'_{(c-3)3} \frac{\Sigma_c D_{c0}}{\Sigma_c D_{(c-3)0}}, \ c = 4, 5, 6$$

$$= (D_{13} + D_{23} + D_{33}) \frac{D_{40} + D_{50} + D_{60}}{D_{10} + D_{20} + D_{30}} .$$

Ex. 6.12. Using equation (6.27) to adjust the numbers of dead of each age. The numbers and conditions of the previous example are used. It is necessary, however, to estimate the expected sample sizes of the D_0 for the previous years. They should be the same as for 1974 and 1975 ($\bar{N} = 30$), except for 1973, which is estimated to be half that, or 15. Results are given to one decimal:

$$D''_0 = (28 + 31)\frac{28 + 31}{28 + 31} = 59.0;$$

$$D''_1 = (4 + 6)\frac{28 + 31}{15 + 28} = 13.7;$$

$$D''_2 = (4 + 1)\frac{28 + 31}{30 + 15} = 6.6;$$

$$D''_3 = (3 + 4)\frac{28 + 31}{30 + 30} = 6.9.$$

The cohort and sampling corrections are stated in general terms throughout this section in order to keep the corrected numbers in life table form. The corrections may often be simplified without introducing error in the calculated rates; for example, a cohort correction equal to $1/(L_{(c-a)0}$ or $1000/(L_{(c-a)0}$ (modified from eq. [6.24]) will not change the estimated rates and may simplify the calculations.

A true current life table is calculated from the L_a present at the start of a period and the D_a dying in the period, and since the latter come from the former, no cohort correction is necessary. This method can be applied directly if the actual numbers are known. Usually, however, only samples of the L_a and D_a are available and a sampling correction is needed. Since the living and the dead normally have quite different probabilities of being sampled, a correction close to 1 cannot be assumed.

Calculation of a Constant Survival Rate from Cohort Data

In many species of animals the survival rate is believed to be constant, or independent of age, in some part of the cohort's history over ages i through k (see section on types of survivorship). In such a case $s_i = s_{i+1} = \ldots = s_k$. The estimate of this constant rate will be symbolized by $\hat{s}.$, the dot to mean "independent of age." The maximum likelihood estimate of $\hat{s}.$ over ages i through k is

$$\hat{s}. = \frac{\Sigma_{a=(i+1)}^{k} L_a}{\Sigma_{a=i}^{k-1} L_a} \tag{6.30}$$

$$= \frac{L_{i+1} + L_{i+2} + \ldots + L_k}{L_i + L_{i+1} + \ldots + L_{k-1}} .$$

Note that the sampling corrections are omitted in the above equation and that they will be omitted in all of this subsection.

The maximum likelihood estimate of the variance of $\hat{s}.$ is

$$\text{Variance of } \hat{s}. = \frac{\hat{s}.(1 - \hat{s}.)}{\Sigma_{a=i}^{k-1} L_a} . \tag{6.31}$$

Ex. 6.13. A constant survival rate from cohort survivorship data. A careful census of an isolated meadow made at intervals of one week recorded the numbers of inhabited spider webs. The numbers in successive weeks were 235, 156, 90, 64, 46. Assuming a closed population, the average survival over the four weeks,

$$\hat{s}. = \frac{156 + 90 + 64 + 46}{235 + 156 + 90 + 64} = .65/\text{week} \qquad \text{(eq. 6.30)}$$

$$\hat{\sigma}_s^{\,2} = \frac{.65 \times .35}{235 + 156 + 90 + 64} = .000417 \qquad \text{(eq. 6.31)}$$

$$\hat{\sigma}_s = .02$$

A constant survival rate means a constant mortality rate, so each $D_a = m.L_a$. Therefore $\hat{s}.$ can also be calculated from

$$\hat{s}. = \frac{\Sigma_{a=(i+1)}^{k} D_a}{\Sigma_{a=i}^{k-1} D_a} ; \tag{6.32}$$

the formula for the estimated variance is parallel to equation (6.31).

Spicer (1956) considered the survival of bacterial populations where an exponential equation, $N_t = N_0 e^{-mt}$ fits the data and m is constant. His maximum likelihood equation for estimating the survival rate for an interval Δt is the same as equation (6.30). The instantaneous mortality rate can then be calculated from $m\Delta t = -ln\ s$. Spicer also provides a maximum likelihood estimate of the variance of m.

Other procedures for calculating a constant survival rate were developed for the analysis of age composition data which are covered in the next subsection. Some of these will be found to apply to cohort data and under certain conditions are superior to equations (6.30 through 6.32).

Calculation of a Constant Survival Rate from Age Composition of Either Living or Dead

Plotting the numbers of fish of each age in a commercial or other catch, using a logarithmic scale on the vertical axis for numbers and a linear scale on the horizontal for age, produces a "catch curve" as illustrated in figure 6.7. Below a certain age the slope of the curve is positive, indicating that as individuals become older they are increasingly likely to be caught. Above this pivotal age the slope is negative

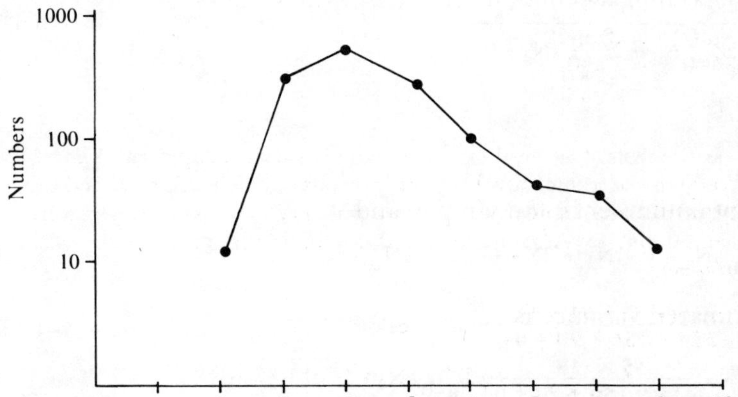

Figure 6.7. A catch curve, numbers caught (logarithmic scale) plotted against age class. Assuming that individuals aged 4 and older are equally likely to be caught, the approximate linearity of the descending limb indicates a survival rate independent of age.

and reasonably constant leading to the interpretation that (a) the probability of being caught is independent of age so that the catch is a sample of the age composition of the living older than the pivotal age, (b) the population is at least approximately stationary, and (c) the survival rate is constant so that the decrease in numbers with age is an exponential decrease. One method of estimating the constant survival rate is by using equation (6.30), using the symbol L_a for the number of each age in the catch. A more precise method is due to Chapman and Robson (1960) who developed it for the analysis of a catch curve (see also Robson and Chapman 1961). They assumed the conditions described above so that neither a cohort correction or a sampling correction was needed. These assumptions will be carried through most of the following in order to simplify the notation. Another and an important condition is that the catch or sample must include the oldest age class; this means that no significant number of older individuals exist in the population than the oldest in the catch, and that the oldest cohort represented in the sample has attained its maximum age.

Under the above conditions the data consist of the number of individuals of each age, N_a, with a ranging from the youngest age class (designated age 0 in the equations following) which fits the assumptions of equal probabilities of capture and equal survival rate to the oldest age. Robson and Chapman (1961) found that the minimum variance, unbiased estimate of the constant survival rate is

$$\hat{s}. = \frac{\Sigma^{\omega}_{a=0} aN_a}{\Sigma^{\omega}_{a=0} N_a + \Sigma^{\omega}_{a=0} aN_a - 1}, \qquad (6.33)$$

where

$$\Sigma^{\omega}_{a=0} N_a$$

is the total number in the sample, and

$$\Sigma^{\omega}_{a=0} aN_a = 0 + N_1 + 2N_2 + \ldots + \omega N_{\omega}.$$

The estimated variance is

$$\text{Variance of } \hat{s}. = \hat{s}. \left(\hat{s}. - \frac{\Sigma^{\omega}_{a=0} aN_a - 1}{\Sigma^{\omega}_{a=0} N_a + \Sigma^{\omega}_{a=0} aN_a - 2}\right), \qquad (6.34)$$

and the large sample variance is

$$\text{Variance of } \hat{s}. = \frac{\hat{s}.(1 - \hat{s}.)^2}{\Sigma^{\omega}_{a=0} N_a}. \qquad (6.35)$$

Equation (6.33) may also be written in terms of the mean age of the animals in the sample. Since the mean age

$$\bar{a} = \frac{\Sigma_{a=0}^{\omega} aN_a}{\Sigma_{a=0}^{\omega} N_a},$$

then

$$\hat{s}. = \frac{\bar{a}}{1 + \bar{a} - \frac{1}{\Sigma_{a=0}^{\omega} Na}}. \tag{6.36}$$

Ex. 6.14. Constant survival rate from a catch curve. The data on ages of fish caught in a commercial fishery are tabulated in the first two columns below. The numbers, when plotted on a logarithmic scale against age, form almost a straight line for ages five and older. If the population is believed to be stationary and if ages five and older are sampled equally by the catching methods, a constant survival rate for these ages is indicated. To apply equation (6.33), the ages are renumbered, 5 becoming 0, and so on.

Age	Number	a	aN_a
2	75		
3	498		
4	724		
5	1150	0	0
6	671	1	671
7	264	2	528
8	61	3	183
9	28	4	112
10	7	5	35
11	0	6	0
Total age 5–11:	2181		1529

$$\hat{s}. = \frac{1529}{2181 + 1529 - 1} = .41224. \quad \text{(eq. 6.33)}$$

$$\hat{\sigma}_s^2 = .41224(.41224 - \frac{1529 - 1}{2181 + 1529 - 2}) = .0000651 \quad \text{(eq. 6.34)}$$

$$\hat{\sigma}_s = .0081.$$

or

$$\hat{\sigma}_s^2 = \frac{.41224(1 - .41224)^2}{2181} = .0000653 \quad \text{(eq. 6.35)}$$

$\hat{\sigma}_s = .0081.$

Mean age of fish in "straight-line" portion of catch = 1529/2181 = .701 year, or

$$\hat{s}. = \frac{.701}{1 + .701 - (1/2181)} = .41222. \quad \text{(eq. 6.36)}$$

In many species of animals whose age is estimated from tooth wear, scale annuli, or a similar characteristic, the older age groups cannot be identified and must be combined in a class containing two or more of the oldest ages. Robson and Chapman (1961) provide a formula for estimating $\hat{s}.$ under these conditions. Let N_v be the number of individuals aged v and older, then

$$\hat{s}. = \frac{\Sigma_{a=0}^{v-1} aN_a + vN_v}{\Sigma_{a=0}^{v-1} N_a + \Sigma_{a=0}^{v-1} aN_a + vN_v}, \quad (6.37)$$

with the variance

$$\text{Variance of } \hat{s} = \frac{\hat{s}.(1 - \hat{s}.)^2}{\Sigma_{a=0}^{v} N_a (1 - \hat{s}.^v)}. \quad (6.38)$$

Ex. 6.15. Constant survival rate from a catch curve with older ages combined. The data from the previous example are used, except for combining ages eight and older, as if these ages could not be accurately determined. Using the symbols of equation (6.37), a = 0, 1, or 2, and v (for ages eight and older) = 3.

a	N_a	aN_a
0	1150	0
1	671	671
2	264	528
	2085	1199
$v = 3$	96	288

$$\hat{s}. = \frac{1199 + 288}{2085 + 1199 + 288} = .416293 \quad \text{(eq. 6.37)}$$

$$\hat{\sigma}_s^2 = \frac{.416293(1 - .416293)^2}{(2085 + 96)(1 - .416293^3)} = .000070 \quad \text{(eq. 6.38)}$$

$\hat{\sigma}_s = .0084$

When the data consist of the number of dead of each age, D_a, including the oldest age class, a constant survival rate can be calculated from the formula first presented by Haldane (1955); this assumes that no cohort or sampling corrections are necessary:

$$\hat{s.} = \frac{\Sigma_{a=0}^{\omega} aD_a}{\Sigma_{a=0}^{\omega} D_a + \Sigma_{a=0}^{\omega} aD_a}, \tag{6.39}$$

which has the large sample variance,

$$\text{Variance of } \hat{s.} = \frac{s.(1 - s.)^2}{\Sigma_{a=0}^{\omega} D_a}. \tag{6.40}$$

The mean age of the dead, which is the average longevity of the individuals in the sample, may also be used in a formula parallel to equation (6.36).

Eberhardt (1972) performed a series of tests on various methods of calculating a constant survival rate and concluded that the Chapman-Robson (equations 6.33 to 6.35), and the parallel Haldane (equations 6.39 and 6.40) methods are clearly the best, with the smallest variance and most robustness. Equation (6.30) gave equally accurate results but with a variance larger by a factor of $1/(1 - \hat{s.})$, which would be expected from comparing equations (6.31) and (6.35).

Chapman and Robson (1960) and Haldane (1955) both give methods for estimating $\hat{s.}$ when the oldest age classes are not included in the sample, and the procedure is also described by Seber (1972). The calculations are trial-and-error, or iterative, and Chapman and Robson provide a table to simplify the work. In his trials of this method Eberhardt (1972) found that the results varied and the method is not robust. In cases of this kind, equations (6.30) and (6.32) are much easier to use and appear to be more reliable.

If the survival rate is constant, the data fit an exponential curve or a geometrical progression and the logarithm of L_a or N_a is a linear function of time or age (see section 6.5). Therefore, linear regression methods can be used and an estimate of $\hat{s.}$ obtained from the slope of the straight line. This method is not recommended, however, as it provides a best fit to the logarithms rather than to the actual numbers, and the few individuals in the older age classes have too much weight in determining the slope.

So far this discussion of a constant survival rate has assumed that neither a cohort nor a sampling correction was necessary. If either or both are required, the procedure is to correct the sample number, either the L'_a, the N'_a, or the D'_a, to the adjusted numbers—L''_a, N''_a, or D_a''—and to use these adjusted values in the equations given above for calculating s.

The assumption of a constant survival rate is most likely to be wrong for the younger age classes. Therefore the survival rate for the youngest group included in the estimate should be calculated separately and compared with the estimated constant rate. If the two are significantly different, the youngest age class should be excluded and the calculations repeated beginning with the next older age group. A test of differences between rates is given in section 6.7 of this chapter.

Survival Rates for Stages of Unequal Duration

Many species of insects have one distinct generation at a time and life history stages (e.g., eggs, larvae, etc.) of unequal duration. Life tables have been constructed for these by intensive sampling of the numbers of each stage and treating the estimated densities as the survivorship of a cohort (Harcourt 1969). If the time of egg-laying or of hatching is spread out somewhat, overlapping of stages results; Southwood (1966, pp. 278–83) describes methods of handling this and related problems. The data are suitable for construction of a life table and calculation of stage-specific rates even though the stages are of unequal duration.

Other species have overlapping generations and eggs being laid repeatedly or continuously, resulting in the development of an age structure with the "ages" being stages of unequal duration. Because the stages present at any one time come from eggs laid at different times, a cohort correction is necessary. A second complication is explained in the following paragraph.

This centers around a hypothetical example of an insect population which has L_0 newly laid eggs, L_1 newly hatched larvae, L_2 larvae hatched 1 week, and L_3 larvae hatched 2 weeks previous to the hatching of the L_1. Each age of larvae came from eggs equal in number to L_0, so no cohort correction is necessary. The egg survival rate is denoted by s_{eg}, and the weekly larval survival rate which is constant by s_{lv}. The age composition would appear as follows:

Eggs:	L_0
New larvae:	$L_1 = L_0 s_{eg}$
One-week larvae:	$L_2 = L_0 s_{eg} s_{lv}$
Two-week larvae:	$L_3 = L_0 s_{eg} s_{lv}^2$

The three ages of larvae are recognized as one stage, so the ratio of larvae present at one time to eggs present can be stated thus: larvae/eggs = $(L_1 + L_2 + L_3)/L_0 = s_{eg}(1 + s_{lv} + s_{lv}^2)$, which is not equal to s_{eg} and which cannot be solved because two unknowns are present.

Therefore the estimation of rates based on age composition is not adaptable to situations where the "ages" are stages of unequal duration. There is, however, an exception to this when the survival rate is believed to be equal in both stages; the procedure is described in chapter 9, the last part of section 9.4.

Using Marked Individuals to Estimate Rates

Several studies have used marked individuals and the data from their recovery to estimate survival rates, especially of birds (Farner 1955, Hickey 1952). The size of each cohort is then determined by the number marked each year. The age at marking is rarely if ever age 0 for the individual being marked, but for convenience it may be considered as coded age 0 in making the calculations. Figure 6.6 illustrates the events and the numbers, with L_{10} being the number marked and released in the first year and the oblique lines of L_{ca} and D_{ca} representing the history of each marked group. Data from the oldest age class will not be available until the first marked cohort has expired, so that some of the estimates described earlier cannot be used until the recoveries have continued long enough for this to occur.

An important source of error in using data from marked individuals is that the marks or tags may be lost; as a result some individuals that were marked will be unrecognized, and the survival rates will be underestimated. Methods of estimating the rate of loss of marks are discussed in chapter 3. Robson and Regier (1966) found that the persistence of tags on fish declined exponentially, or in other words the tags had a constant survival rate of their own. If this can be measured, a correction can be made based on the following logic. Let t be the tag survival rate, and let age, a, be measured from the time of tagging. Then P(tagged individual surviving to age a and keeping tag) = $l_a t^a$. The observed ratio of

$$\frac{\text{tagged individuals of age (a+1)}}{\text{tagged individuals of age a}} =$$

$$\frac{l_{a+1}\, t^{a+1}}{l_a\, t^a} = s_a t. \qquad (6.41)$$

Therefore the observed ratio of apparent survivors should be divided by the tag survival rate to obtain the population survival rate.

Hickey (1952, table 5) presents data showing that the aluminum bands put on waterfowl become illegible at rates varying with age.

The "illegibility rate" appears to be low for the first three years after the band has been attached, then to rise rapidly through years 4, 5, and 6, and then more slowly thereafter. Kadlec and Drury (1968) and Ludwig (1967) found very similar rates of band loss by gulls. The conservative approach to this particular case is to use data only from bands less than four years old. The difficult approach is to find ways of estimating band loss and of making corrections. Perhaps the practical solution is to use all the available data recognizing the source of error and the bias which it introduces into the estimates. Hickey (1952) discusses several other problems in using data from banded birds to estimate survival rates.

Estimating a constant survival rate from returns of marked individuals is usually complicated by the incompleteness of some cohorts, resulting from the fact that marking and recovery operations are limited in time. The problem can be explained most easily by referring to figure 6.6. If the L_{c0} in the diagram are the number marked each year from c equals 1 through 7, and the recovery operations cover the same period, there has been time for only the first four cohorts to be completed. Within these cohorts the recovered L_{ca} or D_{ca} could be summed for each age without any problem (assuming no sampling corrections) and the sums used as if from a single cohort. If, however, one wishes to use all the recoveries, corrections are necessary because the younger ages are from additional cohorts. The cohort correction for this case can be written as

$$(cc) = \frac{\text{(corrected cohort size)}}{\Sigma_{c=1}^{n} L_{c0}}, \qquad (6.42)$$

where n is the number of cohorts from which the recoveries came. The corrected cohort size or base is arbitrary; it could be the sum of the L_{c0} over all cohorts, or a round number like 1000.

Choosing an example from figure 6.6 and an arbitrary base of 1000, the adjusted total of recoveries of living of age two would be

$$L_2'' = \Sigma_{c=1}^{5} L_{c2}' \,(1000/\Sigma_{c=1}^{5} L_{c0}).$$

From the series of adjusted L_a'' a constant survival rate can be estimated by equation (6.30) or (6.33). The same cohort correction (6.42) can be used to correct the sums of the D_{ca}' to D_{ca}'', and the series of these can be used in equations (6.32) or (6.39).

6.7 TESTS OF SIGNIFICANCE BETWEEN ESTIMATED RATES

Mortality and survival rates are proportions, so two rates may be tested for a significant difference by the principle that the difference between two sample proportions taken from one population is normally distributed. More explicitly, if p_1 is the proportion (of any characteristic) in a sample of size N_1, and p_2 is the proportion in another sample of size N_2, both samples having been taken from a single population in which the proportion is $p.$ (which can be estimated by combining both samples), the following quantity has the standard normal distribution:

$$\frac{(p_1 - p_2)}{\sqrt{p.(1 - p.)(\frac{1}{N_1} + \frac{1}{N_2})}} . \tag{6.43}$$

Since the null hypothesis is that $p_1 = p_2$, this would be rejected if the above quantity exceeded the chosen level of probability in either the positive or negative direction.

This test can be applied by substituting the appropriate rates for p_1, p_2, and $p.$, and sample sizes for N_1 and N_2. Some examples are presented in table 6.4. It is important to note that in testing rates calculated from the corrected or adjusted number of survivors (e.g., L''_{a1}), the size of the sample is the observed number, L'_{a1}.

Table 6.4. Quantities to be Substituted in Equation (6.43) for Testing Difference between Rates.

	p_1	p_2	$p.$	N_1	N_2
Mortality rate	$\frac{D_1}{L_1}$	$\frac{D_2}{L_2}$	$\frac{D_1 + D_2}{L_1 + L_2}$	L_1	L_2
Survival rate from observed L_a	$\frac{L_{(a+1)1}}{L_{a1}}$	$\frac{L_{(a+1)2}}{L_{a2}}$	$\frac{L_{(a+1)1} + L_{(a+1)2}}{L_{a1} + L_{a2}}$	L_{a1}	L_{a2}
Survival rate from corrected samples	$\frac{L''_{(a+1)1}}{L''_{a1}}$	$\frac{L''_{(a+1)2}}{L''_{a2}}$	$\frac{L''_{(a+1)1} + L''_{(a+1)2}}{L''_{a1} + L''_{a2}}$	L'_{a1}	L'_{a2}

When constant or independent-of-age mortality rates have been calculated for two cohorts or groups, these can be compared by a test proposed by Farris (1968). Under the condition of a constant mortality rate the survivors of a cohort decline in number exponentially, $L_a = L_0 e^{-ma}$, and the average longevity equals $1/m$. Farris used these properties to show the following. Let the average longevity

Ex. 6.16. Testing survival rates from two habitats. The number of barnacles in a sample area on pilings was 86 at the start of a study period, and 51 survivors were present one year later. In a sample area on rock the numbers were 123 and 59. A check of the accuracy of the counts was made by counting barnacles in other plots, then scraping all barnacles off and counting the separate individuals. The first counts apparently counted only 92 percent of those actually present, while the second counts were accurate because the survivors were larger.

On piling:

$$L'_0 = 86;\ L''_0 = \frac{86}{.92} = 93;\ L'_1 = 51.$$

On rock:

$$L'_0 = 123;\ L''_0 = \frac{123}{.92} = 134;\ L'_1 = 59$$

$$\hat{p}_1 = \frac{51}{93} = .548;\ \hat{p}_2 = \frac{59}{134} = .440;\ \hat{p}. = \frac{51 + 59}{93 + 134} = .485$$

$$t = \frac{(.548 - .440)}{\sqrt{.485(1 - .485)(1/86 + 1/123)}} = 1.537$$

Under the null hypothesis that $p_1 = p_2$, the probability that t will exceed the absolute value of 1.537 is greater than 5 percent, so the null hypothesis would be accepted; that is, the rates are not significantly different from each other.

or mean age at death be $\bar{a}_1$ in a sample of D_1 dead individuals, and $\bar{a}_2$ and D_2 be the same for a second sample, with $\bar{a}_1$ greater than $\bar{a}_2$. Under these conditions, $\bar{a}_1/\bar{a}_2$ has the F two-tailed distribution with $2D_1$ and $2D_2$ degrees of freedom. If the ratio exceeds the values tabulated at the chosen level of significance in a table of the two-tailed F distribution, population or cohort 1 has a significantly greater average longevity and a lower mortality rate. Farris (1968) also describes how confidence intervals can be calculated for the mean age at death, and he provides examples of the calculations.

An area of biological research which involves mortality rates is that of bioassay, where the quantitative effects of poisons, pollutants, radiation, and so forth, are measured by the mortality rate. In the usual procedure the dose or concentration is transformed into its logarithmic value, which is termed the dosage. The proportion of the exposed animals dying within a predetermined period of time (i.e., the population mortality rate) is typically a sigmoid function of the dosage. The slope of the curve is steepest where the mortality rate equals 50 percent, making this point useful for measuring the effect of a treatment and for comparing different treatments. The dosage at this point is termed the LD_{50}, the lethal dose to 50 percent of the sample in the experimental period. The length of the period is, of course, very important and should be stated (e.g., the LD_{50}, 7 days). The estimation of this point, or of the lethal dose for any other mortality rate, and the making of comparisons between different treatments are facilitated by a procedure called probit analysis which is fully described in a book by D. J. Finney (1962).

6.8 LONGEVITY AND TURNOVER

The average longevity is the average age at death and is calculated from a sample of the ages at death by

$$\bar{a} = \frac{\Sigma_a \, aD_a}{\Sigma_a D_a} \, . \tag{6.44}$$

It is also the expectation of life at birth, calculated by life table methods (section 6.3) by

$$\bar{e}_0 = \frac{\Sigma_{a=0}^{\omega} \bar{L}_a}{L_0} \, . \tag{6.45}$$

It was shown in section 6.4 that in a stationary population $\bar{e}_0$ is the reciprocal of the death rate and, under some conditions, also of the mortality rate. The latter can also be shown to be true when the mortality rate is independent of age, whether or not the population is stationary. In at least two cases, then, $\bar{e}_0 = 1/m$.

The concepts of "turnover" and "turnover period" are often used by wildlife managers. The definition of turnover period given by Petrides (1949) is that it is the length of time required for the disappearance of a cohort. This is practically equivalent to maximum longevity which has little significance in population dynamics. A more reasonable definition, based on the use of the term in chemistry, is that the turnover period is the time required to lose from the population the number equal to the size of the equilibrium population. It is easy to prove that by this definition the turnover period is the reciprocal of the mortality rate and equals the average longevity.

6.9 SUMMARY OF METHODS

General survival and death rates can be calculated from counts of individuals whose ages are unknown or are ignored. For the estimation of age-specific rates, data from individuals of known age are required. Such data are often analyzed to determine a constant rate, one independent of age; this is also a general rate for the ages represented in the data. Up to this point the general and constant rates have been kept separate because of different methods of calculation, but they should be recognized as measuring the same rate, one independent of age.

Numbers of dead from each cause can be used to calculate death rates by cause, some relations of these rates are expressed in equations (6.11) and (6.12).

Changes in sex or age ratios may enable estimation of general mortality rates (equations 4.11 through 4.13).

Mark and recapture data have several applications. Equation (6.6) is a simple estimate of survival. A mark and recapture procedure specifically designed to estimate survival rate is described with equation (6.7). A somewhat similar method can be used to estimate mortality due to fishing versus natural mortality (equation 6.14). The multiple mark-recapture method due to Jolly yields estimates of the general survival rate (equations 3.11, 3.15, 3.19). Recoveries of marked, dead animals, under the assumption of survival independent of age, can provide a general survival rate for each period in which the dead animals are recovered, by use of equation (6.9). A correction for the loss of marks is described with equation (6.41).

If data are missing, rates may sometimes be estimated with equation (6.19) or by the method illustrated in example (6.4).

The methods summarized to this point generally assume that all ages are equally likely to be sampled and that the estimates are not biased by changes in population size. In those that follow one or both of these assumptions may not hold and corrections may be necessary. Adjustments for unequal catchability are sampling corrections (equations [6.16], [6.17], [6.18], [6.20]). If the population is not stationary, some methods will require cohort corrections (equations [6.24], [6.25], [6.26], [6.28]).

The history of a cohort can be determined by marked individuals or by some means of determining individual age. A life table can be constructed if the record is complete. Cohorts may sometimes be combined to give larger numbers. These data can be used for measuring age-specific rates (equations [6.16], [6.21] through [6.23]), or for estimating general—that is, independent of age—rates (equations [6.30 through 6.40]).

The age composition of the living or dead (age at death) may be determined in a number of ways: age markers, tagged individuals, catch curves, etc. If the population is assumed to be stationary, both general and age specific rates can be calculated (section 6.4). Information on the age composition of non-stationary populations can be used for the calculation of either age-specific rates (equations [6.24] through [6.29]) or the general survival rate (equations [6.33] through [6.40]).

Differences between estimated rates may be tested for statistical significance by two means presented in section 7.

The reasonableness of any estimated death or survival rate must often be evaluated subjectively, from familiarity with the species or from knowledge of its life history characteristics. As an example of the latter, an estimate of the mortality rate may be matched with the birth rate and growth of the population; methods for doing this are described in chapter 9.

7 Reproduction, Migration, and Recruitment

The individuals added to a population are the newborn and the older ones that move into the area from elsewhere. Movement or migration may of course also remove individuals from the population. This chapter describes the meaning of reproductive and migration rates, and of the process called recruitment which is of interest to fishery and game managers.

7.1 REPRODUCTION

Fecundity is the potential reproduction of offspring, the maximum possible. The tremendous range of fecundity in the biological world is illustrated by the difference between spore-producing fungi and man (whose burgeoning population results not from a high reproductive rate but from an excellent survival rate). Fecundity depends upon a number of physiological characteristics such as the frequency of reproduction, the number produced in each reproduction, and the changes of these with age. Information on many of these characteristics for a variety of species may be found in the reference work edited by Spector (1956). More information on fishes is summarized by Bagenal (1967) and on small mammals by French, Stoddart, and Bobek (1975). Methods of measuring fecundity in aquatic invertebrates are described by Edmondson and Winberg (1971) and in insects by Southwood (1966).

Fertility is the actual reproductive performance of a population, measured by a reproductive rate. Synonyms are birth rate and natality rate. All reproductive rates are a ratio between the number of offspring produced in an interval of time and the number in all or some part of the population. There are questions, however, about the meaning of the words "offspring" and "reproduced." For sexual species, the number of offspring is the number of zygotes formed by fertilization, but usually this is almost impossible to determine. A variety of other measures are used: number of eggs laid, embryos released, eggs hatched, young born alive, and so forth. Asexual reproduction, especially fission, would require different measures. The point is that the decision as to which stage of the life history shall be counted as offspring is arbitrary, and is largely made on the basis of which can be most easily and accurately counted. This is in contrast to definitions about mortality, where there is usually little question about the meaning of "dead" or "survived."

In human populations the annual reproductive or birth rate is the ratio of the young born alive in a year, *B*, to the average number alive in the population during the year, $\overline{N}$:

$$b = \frac{B}{\overline{N}} \tag{7.1}$$

The newborn are included in the calculation of $\overline{N}$, which may be estimated in different ways, as described in chapter 5. The ratio is usually expressed with a base of 1000 or 10,000 of the population.

Equation (7.1) can be used to measure the fertility or the *general reproductive rate* of any population which reproduces continually throughout the year. It is parallel to the general population death rate. Equation (7.1) equals the instantaneous birth rate if this has been constant throughout the interval and if $\overline{N}$ is measured as described in the discussion pertaining to Figure 5.2.

A birth rate measured with respect to the number alive at the beginning of the interval is often referred to as a *finite birth rate*. This rate, β, and a constant instantaneous birth rate, b, are related by

$$b = \frac{r\beta}{(e^r - 1)} \tag{7.2}$$

where r is the population growth rate, equalling $(b - d)$. If the death rate, d, is zero, this equation simplifies to $b = ln(1 + \beta)$. The derivation and uses of these two equations are discussed by Caswell (1972).

Many species of animals have a short annual breeding season when all the offspring of that year appear almost simultaneously. The practical way to measure fertility in this case is by $b = B_t / A_t$ where B_t is the number produced at time t and A_t is the number of adults, animals approximately one year old or older, alive at the start of that reproductive season. This applies also to once-in-a-lifetime breeders such as many species of insects.

An intermediate pattern of reproduction is illustrated by many rodents and small birds that produce two or more litters or clutches in a breeding season lasting up to several months, and then cease reproduction for the remainder of the year. Fertility of these species has often been measured as in the previous paragraph, with A_t equal to the number of adults alive at the start of the season. This does not allow, however, for variable mortality of adults during the season, nor for the early maturity of some rodents which may breed in the same season in which born. A better measure in these cases is to use equation (7.1), basing $\bar{N}$ on either adults only or on the entire population including the newborn, whichever is most significant. In either case the average would be calculated over the period of reproduction to yield a $B/\bar{N}$ per breeding season.

The reproductive rate is usually measured directly using data on the number produced in a time interval and the size of the population. Another method has often been used for those aquatic invertebrates (Edmondson and Winberg 1971) which carry the eggs within or on the body until at the end of a developmental period the eggs hatch and the young are "born." This method can also be applied to any species with a similar reproductive pattern, including viviparous vertebrates. The data needed are the duration of the developmental period and the egg or embryo ratio, the average number carried per adult; this information is usually obtained from a sample. Let D be the length of the developmental period and $\bar{E}$ be the average egg or embryo ratio, then the instantaneous rate can be calculated from a formula derived by Paloheimo (1974):

$$b = \frac{\ln(\bar{E} + 1)}{D} \tag{7.3}$$

The average egg ratio divided by the duration of the developmental period, $\bar{E}/D$, has often been used as an estimate of the finite birth rate, but Paloheimo (1974) shows that this is correct only under certain circumstances.

Ex. 7.1. Birth rate from egg ratio. A population of a cladoceran sampled in midspring had an average egg ratio of .8 egg per female. The eggs are carried by the female for 7 days, at which age they hatch. The instantaneous birth rate from equation (7.3) was

$$b = \frac{\ln(.8 + 1)}{7}$$

$$= .084/\text{female/day}.$$

An estimate of the finite daily birth rate is desired. The population growth rate is unknown, but an upper limit for the finite birth rate can be estimated by assuming that the death rate of adult females during the developmental period is zero, i.e., the growth rate equals the finite birth rate. Then, modifying equation (7.2) for this assumption,

$$\beta = e^{b} - 1 = e^{.084} - 1$$
$$= .088/\text{female/day}$$

The egg or embryo ratio is usually estimated for females only, so the resulting birth rate is sex specific. Another specific reproductive rate that is very useful in population studies is termed the age-specific maternity rate; it counts only the number of female offspring born (for example, in one year) to mothers of each age. Let B_{fa} be the number of daughters born to mothers of age a, and $\bar{N}_{fa}$ be the average number of females of age a alive during the period, then the *age-specific maternity rate* is

$$b_{fa} = \frac{B_{fa}}{\bar{N}_{fa}} \ . \tag{7.4}$$

(In many works b_{fa} is symbolized by m_a, but not here because m has been preempted for the mortality rate.)

When the age-specific maternity rate, b_{fa}, is multiplied by the probability of a female living to age a, l_{fa}, and these products are summed over all ages, the total is the number of daughters expected to be born to a female just born. This is termed the *net reproductive rate*,

$$R_0 = \Sigma_a l_{fa} b_{fa} \ . \tag{7.5}$$

Despite the name, R_0 is not a true population rate; it is better called a *generation ratio*, since it is the ratio of female births in two successive generations.

Ex. 7.2. Calculation of a net reproductive rate or generation ratio. An insect summer generation has the following set of l_a and b_{fa}. Age is measured in weeks. The b_{fa} equals one half of the number of eggs laid per female in each of the three weeks in which eggs are laid.

a	l_a	b_{fa}
0	1.0	0
⋮	⋮	⋮
5	.32	0
6	.25	30
7	.20	45
8	.10	15
9	0	-

That is,

$$R_0 = (.25 \times 30) + (.20 \times 45) + (.10 \times 15) = 18. \qquad \text{(eq. 7.5)}$$

Other measures of fertility used in demography are the "general fertility rate," "total fertility rate," and "gross reproductive rate"; definitions of these can be found in demography texts (e.g., Keyfitz and Flieger 1971, Spiegelman 1968). They have little application in ecological studies.

7.2 DISPERSAL AND MIGRATION

Dispersal refers to the moving or scattering of individuals, usually from their place of origin. It may be passive, caused by winds or currents, or it may result from more or less random wandering. In birds and mammals the immature individuals are usually the ones that disperse, but in insects the adults are generally more mobile. Methods for the study of dispersal by means of marked animals are summarized in Southwood (1966) and Poole (1974); the examples given in these references are of insect dispersal.

One result of dispersal is that individuals leave one population and enter another, thus changing the population densities; this is termed migration. (The periodic movement of whole populations from one area to another is also called migration, but that is another subject.)

Emigration, or out-migration, is measured by the ratio (number of emigrants)/(average population size) in an interval. Thus the migrants are related to the size of the population exposed to the risk of migration. This is exactly comparable to the death rate, and the two rates combined measure the rate at which individuals are lost from a population. In most studies of wild populations it is impossible to learn whether an individual has disappeared by death or by emigration and the different events are combined into one measure, or are measured by the complementary rate, the survival rate. This is the meaning of survival rate in the multiple mark-recapture method.

An immigration rate cannot be so clearly defined because the immigrants are not produced by the receiving population, and there is no functional relation between them and the size of this population. In the extreme case, when a species colonizes a new area, the population is zero until immigration occurs. Two measures of immigration are therefore used. One is the absolute immigration rate: number of immigrants per time. The second is the population immigration rate, similar to the emigration rate given above, and is required when the net migration rate, the difference between the two, is desired: net migration rate = (immigrants - emigrants)/(average population size) in an interval.

7.3 RECRUITMENT

Recruitment is the process by which individuals are added to a certain size or age class in a population. It is especially important in fisheries, where it means the addition to the catchable portion of the population. In this connection, Ricker (1975) defines the rate of recruitment as the number growing to (or being added to) the catchable size per period of time divided by the number of catchable fish already present at that time. He uses the symbol z for this rate. Recruitment is also important in the management of game species.

Although the term is usually applied to the catchable, vulnerable, or usable portion of a population, it may be used for any age group or other segment. For example, the reproductive rate may be considered as a special case: the rate of recruitment to the zero age class.

Recruitment to age groups older than the newborn results from both growth and immigration; the number added is the sum of the individuals already in the population that grow or mature to the necessary condition plus those of that condition that immigrate. If the population is closed (no immigration), the amount of recruitment is determined by the amount of reproduction that occurred some time in the past and the survival rates since then. The steps can be visualized with the aid of figure 6.6. If k is the age at which individuals mature into the catchable stock, the number of these at time t are the survivors of a cohort produced $(t - k)$ years ago; that is, $L_{(t-k)k} = L_{(t-k)0}l_k$. The newborn of cohort $(t - k)$ are the product of that year's population times the general birth rate, b, of that year. And finally, the recruitment rate at time t will be $z = bl_k$(population at time $(t - k)$)/(catchable stock at time t). Recruitment is therefore not as directly a function of the size of the immediate population as is reproduction. If migration also has an effect on recruitment, there is an additional reason for recruitment rates varying more than do reproductive rates.

In chapter 3, section 3.4, on the multiple mark-recapture method of estimating population size, the number recruited, B_i, is the number added to the catchable portion of the population by any means. Equation (3.12) is an estimation of this number. Chapter 4 describes methods (equations [4.14] and [4.15] of measuring recruitment from changes in age or sex ratio.

Because of the importance of recruitment in fisheries, Ricker (1975) discusses the process in relation to other subjects, including its effects on the accuracy of population estimates and on the interpretation of catch curves.

8 Measurement and Analysis of Population Growth Rate

This chapter describes measurement of observed increases or decreases in population size and methods of analyzing these changes. For convenience, the word "growth" will be used for both increases and decreases in numbers. Chapter 9 is on the mathematical relations between growth and other population parameters, and chapter 10 is an introduction to modeling or predicting population changes.

8.1 GROWTH RATE AND RELATED MEASURES

The *instantaneous growth rate* may be defined in the two equivalent expressions:

$$r = \frac{1}{N}\frac{\delta N}{\delta t},$$

or

$$\frac{\delta N}{\delta t} = rN. \qquad (8.1)$$

If r is constant from time zero to time t, the number will rise exponentially to

$$N_t = N_0 e^{rt}, \qquad (8.2)$$

or will decrease exponentially if r is negative. When defined and used in this way for a closed population, r is occasionally called the "crude rate of natural increase."

An *average growth rate* is measured over a period, Δt. When births, deaths, and migration occur continually throughout the period, as is normal for human populations, the change is related to the average number alive during the period. Over an interval Δt,

$$r = \frac{\Delta N}{\bar{N}\Delta t}\ , \qquad (8.3)$$

$$= \frac{B - D + I}{\bar{N}\Delta t} = b - d + i\ ,$$

where *I* and *i* refer respectively to net migration and the net migration rate.

A relation between the average and instantaneous growth rates can be developed as follows, with the help of figure 8.1. Assume that during a period Δt the instantaneous growth rate varies as shown by

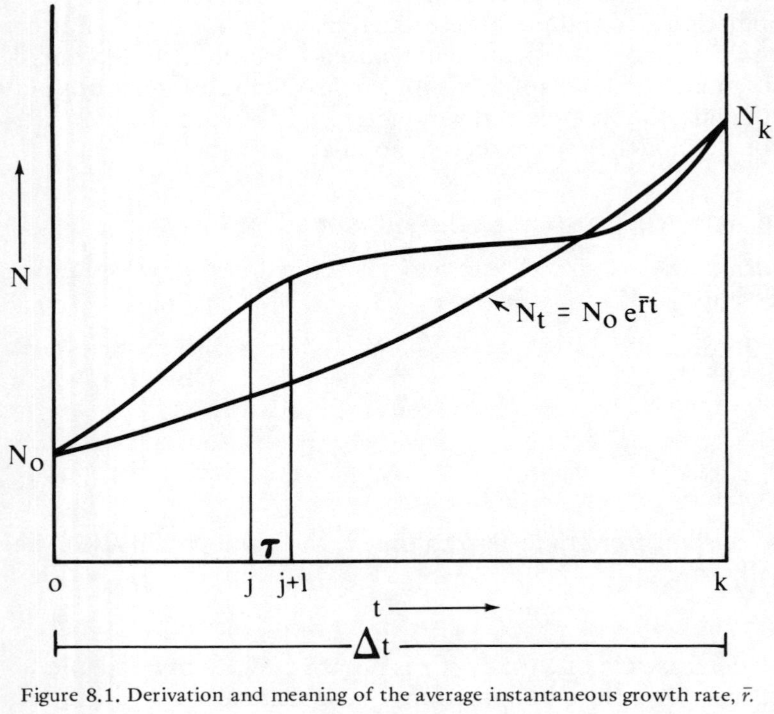

Figure 8.1. Derivation and meaning of the average instantaneous growth rate, $\bar{r}$.

the fluctuating curve of the figure. Divide the time interval Δt into k subintervals, each of length τ, such that over each subinterval the instantaneous rate is constant and therefore $N_{j+1} = N_j e^{r\tau}$. From this, $r_j\tau = ln N_{j+1} - ln N_j$. Summing this equation over all k subintervals, and replacing τ by $\Delta t/k$, $(\Delta t/k)\Sigma r_j = \bar{r}\Delta t = lnN_k - lnN_0$, and therefore the number at the end of Δt, $N_k = N_0 e^{\bar{r}\Delta t}$. Following the same argument as used in chapter 5, if $\bar{N}\Delta t$ equals the area under the curve of growth—that is if it equals the total time lived in the interval—then the average growth rate as defined by equation (8.3) equals the average instantaneous growth rate, $\bar{r}$. This equality is of little practical importance, but it is occasionally useful in the mathematics of population growth.

The direct way to measure the growth rate over a period is to use equation (8.3) and information on the change in numbers and the average population size over the interval. A second method is to transform equation (8.2) to $ln N_t = ln N_0 + rt$, and to calculate r from this. If several estimates of N at different times have been made and if r is assumed constant, $ln N_t$ can be treated as a linear function of t and linear regression methods can be used to estimate the regression coefficient, which is r.

Most species of animals do not reproduce continually, but have a short reproductive season during which their populations increase sharply, followed by a longer period of decline until the next breeding season. A different base for measuring growth is required. Over a single cycle, the growth rate (symbolized in this case by r') will be

$$r' = \frac{\Delta N}{N_0} = \frac{N_1 - N_0}{N_0} \,. \qquad (8.4)$$

N_0 is the number in the population at any arbitrarily chosen point in the cycle. Most convenient is the number in the population just before the reproductive season, but for some species such as insect pests the number of hatching larvae may be more significant. In the next chapter it will be shown that in a stable population (with constant birth and death rates) the growth rate would be the same no matter which point in the cycle is chosen.

Growth Ratio

A second method of measuring or stating a population's growth is by the ratio of its numbers at successive times, the *growth ratio*:

$$g = \frac{N_1}{N_0} = \frac{N_{t+1}}{N_t} \,. \qquad (8.5)$$

This is often called the "finite rate of increase," a poor term, because it is not a rate in the usual sense and also because the population may be decreasing. A special case of this ratio is the net reproductive rate, R_0 of equation (7.5), which is the ratio of female births in two successive generations.

The growth ratio produces geometric growth: $N_x = N_0 g^x$, $x = 0, 1, 2, \ldots$

The growth ratio may be determined directly from the ratio of the number in the population at two successive times. If repeated measurements of N have been made at the same points in the cycle, g may be calculated by transforming $N_x = N_0 e^{gx}$ to $\ln N_x = \ln N_0 + x(\ln g)$, and using linear regression methods to estimate the regression coefficient, $(\ln g)$.

Doubling Time

A third way to state the growth of a population is by the time required for the population to double if it is increasing, or to be halved if decreasing. This measurement is not particularly useful in analytical studies, but it is often an effective way to grasp the implications of growth; for example, there is considerable impact in the statement that today the doubling time of some Latin American countries is 20 years as compared with 100 years for European countries (Keyfitz and Flieger 1971). Doubling time, T_D, can be derived from $2N_0 = N_0 e^{rT_D}$, from which $T_D = \ln 2/r = .693/r$.

Comparisons

Relations between the four measures of population growth are shown below. The growth rate, r, is as defined in equations (8.1) or (8.3):

$$r' = e^r - 1;$$

$$g = e^r = 1 + r'; \qquad (8.6)$$

$$T_D = .693/r.$$

They are compared graphically in figure 8.2. This shows how they change with increasing or decreasing growth, and how the values are related at two critical points: when the population is stationary and when it becomes extinct.

Ex. 8.1. Measurements of population growth. A population increases in one year from $N_0 = 1000$ to $N_1 = 1200$; $\bar{N} = 1100$, approximately.

$$r = \ln \frac{1200}{1000} = .182 \qquad \text{(eq. 8.2)}$$

$$r = \frac{200}{1100} = .182 \qquad \text{(eq. 8.3)}$$

$$r' = \frac{200}{1000} = .200 \qquad \text{(eq. 8.4)*}$$

*Both r and r' would usually not be calculated for the same case.

$$g = \frac{1200}{1000} = 1.200 \quad \text{(eq. 8.5)}$$

$$T_D = \frac{.693}{.182} = 3.81 \text{ years} \quad \text{(eq. 8.6)}$$

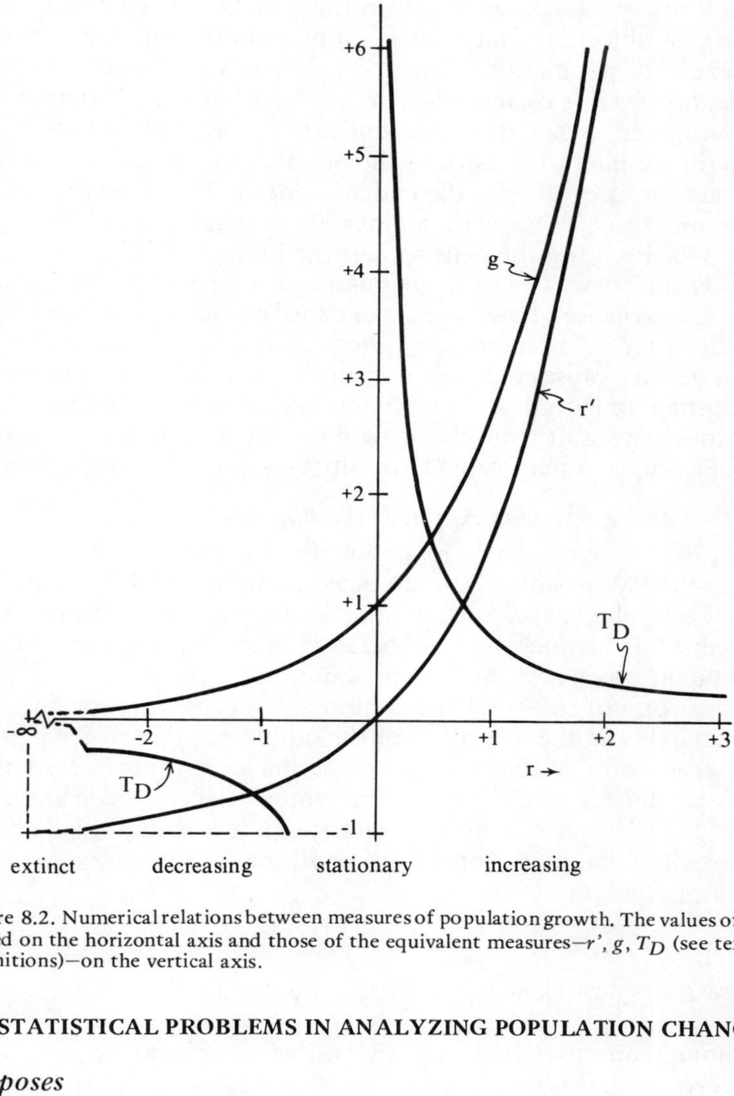

Figure 8.2. Numerical relations between measures of population growth. The values of r are scaled on the horizontal axis and those of the equivalent measures—r', g, T_D (see text for definitions)—on the vertical axis.

8.2 STATISTICAL PROBLEMS IN ANALYZING POPULATION CHANGES

Purposes

One objective of the analysis of a population record is to investigate how populations may be regulated by testing the hypothesis that the population growth rate is some function of the population density. Another is to find means of managing a population by identifying the key processes which determine numbers. A third objective of some research is to learn which parameters are the best predictors of future population size.

Many, if not most, of the studies in this area have been concerned with the first objective, that of attempting to determine the "density-dependence" of populations. This term is a source of confusion. Its converse, however, is clear; a growth rate (or whatever parameter is being considered) is "density-independent" if it is random with respect to the number or density of the population. The first term is ambiguous because "density-dependent" does not state whether a rate increases or decreases with an increase in density. In either case the rate depends upon the density, but the biological effects would be entirely different. The following usages are clear, even though not satisfactorily concise; if an increase in density results in an increase in a rate, the latter is "an increasing function of density"; but if an increase in density causes a decrease in a rate, or a decrease causes an increase, the rate is "a decreasing function of density." Stability in populations can result from the growth rate being a decreasing function of density, or perhaps from mortality being an increasing function.

Correlation and Regression Applied to a Population Record

Morris (1963) developed a regression method which came to be known as the "Morris plot" (summarized in Southwood 1966, pp. 298–302). His objective was to predict future population numbers of pest insects, but the method was also used in attempts to detect density-dependent processes which control numbers. The plot is a graph of the logarithm of the number in one stage or generation on the horizontal axis and the logarithm of the number in the succeeding stage or generation on the vertical axis. Using logarithms instead of numbers produces two advantages: an exponential change is linearized, and the variance around the regression line is stabilized. The points usually fall on a reasonably straight line, so that a regression equation can be written:

$$\ln N_{i+1} = a + b \ln N_i . \tag{8.7}$$

The regression coefficient b is estimated by the usual statistical procedures.

Combining equations (8.2) and (8.7), the growth rate, $r = ln(N_{i+1}/N_i) = (a + b\ ln\ N_i) - ln\ N_i = a + (b - 1)\ ln\ N_i$. If b, the regression coefficient of equation (8.7) equals 1, then $(b - 1) = 0$, and r is independent of density. If b is less than 1, the growth rate is a decreasing function of density, and vice versa.

Other students of this problem, including myself (Tanner 1966), have used correlation measures on the numbers directly (not on logarithms). All of the methods so far developed, however, including the Morris plot, are inadequate for analyzing a population record. The root of the problem is that a population record is a "time series," a sequence of measurements of one variable made at successive times. In the analysis of such a series there are several statistical problems which have been reviewed by Ito (1972). The major one is that a correlation or regression of this nature, a so-called autoregression, is biased (Poole 1972, Reeves 1972, St. Amant 1970). The nature of the bias can be stated as follows: if $\beta = 1$ (the true regression coefficient is 1), the maximum likelihood estimate, b, will be less than 1 if there is any random variation in the data. Therefore, if there actually is density independence, a correlation or regression analysis of a population record will probably indicate b less than 1 suggesting that the growth rate is a decreasing function of density.

Varley and Gradwell (1968) have suggested a method for evaluating the bias mentioned above; its basis may be described as follows. If a population record varies randomly (density-independent changes), the regression $ln\ N_{i+1} = a + b\ ln\ N_i$ will yield a value of b less than 1. Also a regression that is reversed, setting $ln\ N_{i+1}$ as the independent variable, $ln\ N_i = a' + b'\ ln\ N_{i+1}$ will yield a value of b' less than 1. As illustrated in Figure 8.3A, the two regression lines drawn on the same

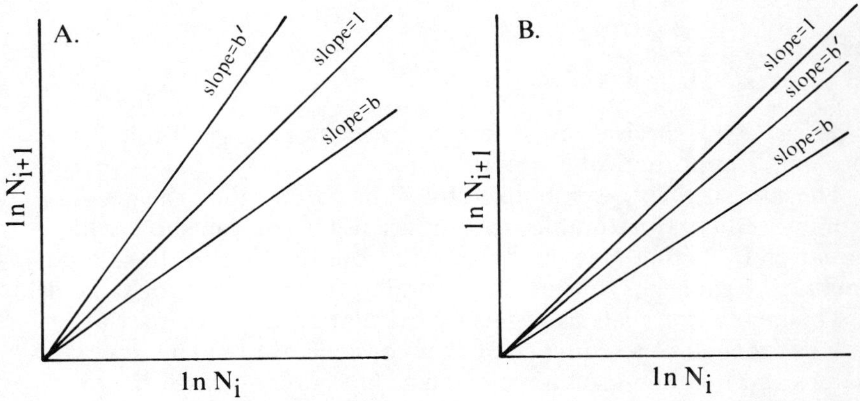

Figure 8.3. Illustrations of Varley and Gradwell's method for assessing bias in regressions of $ln\ N_{i+1}$ on $ln\ N_i$. (A) Regressions for the density-independent case. (B) Regressions for the growth being a decreasing function of density.

graph will lie on opposite sides of the line with slope 1 (fig. 8.3A). On the other hand, if b is less and b' is greater than 1, both regression lines will lie on the same side of the line with slope 1 (fig. 8.3B). From this we may conclude that the density-dependence is great enough to overcome the bias resulting from random errors or variations, and therefore that the growth rate is a decreasing function of density. With b greater than 1 and b' less than 1, the opposite conclusion would be reached.

The above described procedure will not lead to a test of significance nor to an estimate of confidence intervals. This discussion may be summarized by saying that at present there is, unfortunately, no satisfactory method of applying correlation or regression analysis to a population record.

k-Value Analysis

This is a method and terminology proposed by Varley and Gradwell (1960) and explained in detail by Varley, Gradwell, and Hassell (1974, pp. 115–25). Its objective is to determine the life history stages in which key factors operate, these being the factors most important in determining the increase or decrease from generation to generation. It has been applied primarily to insects with discrete generations.

The data consist of the numbers of each successive stage in a generation. From the logarithms of each pair of successive numbers is calculated a "k-value,"

$$k_i = \log N_i - \log N_{i+1}, \qquad (8.8)$$

$$= \log \frac{N_i}{N_{i+1}} = \log \frac{1}{s_i},$$

where s_i is the survival rate of the N_i to the next stage. Table 8.1 presents a hypothetical example.

The sum of the k_i, symbolized by K, has been called the generation mortality. If natural logarithms are used, you can show with equation (6.4) that K is the death rate from the first to the last stage included (from eggs to time of reproduction, in the case of table 8.1).

The key-factor analysis begins by calculating as above a set of k-values for each of a sequence of years or generations. The separate k_i values and K values are then plotted as in figure 8.4, with the years (or generations) on the horizontal axis and scales for the k-values arranged on the vertical axis so as to show clearly the year-to-year differences in each. The k-value which makes the largest contribution to the generation K is then determined by inspection or by a statistical method (Podoler and Rogers 1975). The k-value so identified may

Table 8.1. Data and Calculations for k- Value Analysis.

Data	Symbol	k-values
Actual or potential eggs laid	N_0	
		$k_0 = \log N_0 - \log N_1$
Eggs hatched	N_1	
		$k_1 = \log N_1 - \log N_2$
Larvae pupating	N_2	
		$k_2 = \log N_2 - \log N_3$
Adults emerging	N_3	
		$k_3 = \log N_3 - \log N_4$
Adults reproducing	N_4	
		$\Sigma k_i = K = \log N_0 - \log N_4$

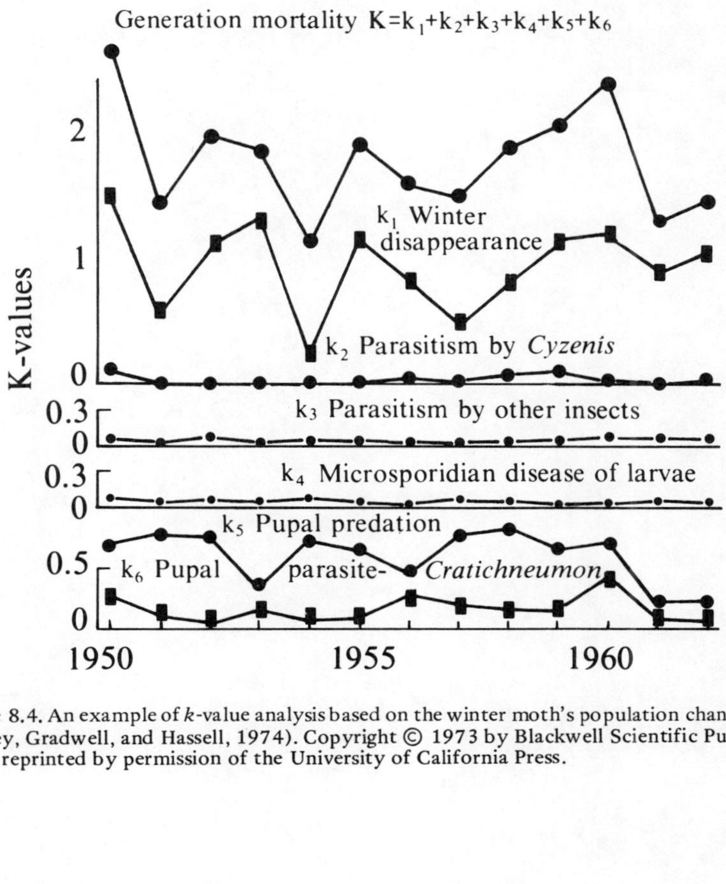

Figure 8.4. An example of k-value analysis based on the winter moth's population changes (Varley, Gradwell, and Hassell, 1974). Copyright © 1973 by Blackwell Scientific Publications; reprinted by permission of the University of California Press.

then lead to the key factor which is largely responsible for population fluctuations. For example, in figure 8.4 changes in K are largely due to changes in k_1, caused by losses over winter.

The question of whether or not the k-values are density-dependent is biologically important. The obvious approach to the question is by the equation $k_i = a + b \log N_i$, and by testing whether or not b is significantly different from zero. By substituting for k_i from equation (8.8) this equation can be changed to the same form as (8.7), and so the analysis would be subject to the same bias as described in the preceding section. An analysis of the bias could be made as described there.

Discussions and examples of k-value analysis are presented by Varley, Gradwell, and Hassell (1974), by Podoler and Rogers (1975), and by Dempster (1975).

9 The Mathematical Relations between Population Characteristics

The interrelations of population characteristics have several practical applications. For example, some population characteristics can be estimated indirectly from knowledge of others (sections 9.3, 9.4, and 9.5). Practical use is often made of the comparison between known and predicted age compositions (example in section 9.3). The reasonableness of estimated survival rates can be checked with information on reproduction and population growth (example in section 9.4). Prediction of population changes for practical or hypothetical reasons can be made from an understanding of population dynamics.

In previous chapters population characteristics have largely been treated separately, except for a brief discussion in chapter 6, section 6.4, on the unique but important case of the stationary population. In this chapter the relations between different characteristics will be discussed on a mathematical basis with the aim of arriving at useful conclusions.

9.1 THE STABLE POPULATION

A closed population is completely described (except for density, dispersion, and sex ratio) by six parameters (Keyfitz 1968): (1) b, the reproductive rate; (2) d, the death rate, or s, the survival rate; (3) r, the growth rate; (4) l_a, the probabilities of surviving to age a; (5) b_a, the age-specific reproductive rates; and (6) C_a, the proportion of the population between ages a and $a + 1$. These are not independent of each other.

A population with constant age-specific survival and reproductive rates will eventually attain an unchanging or stable age composition that is determined only by these rates (Lotka 1922). Costello and Taylor (1971) provide a proof of this conclusion. Such a population, termed a stable population, has the following characteristics:

1. r is constant (not necessarily zero);
2. $N._t = N._0 e^{rt}$;
3. $B_t = B_0 e^{rt}$; (9.1)
4. $N_{at} = N_{a0} e^{rt}$. (9.2)

$N.$ refers to the total number, B to the newborn, N_a to the number of age a, and the subscripts 0 or t indicate the beginning or end respectively of the time interval over which the changes occurred. Equation (9.2) may be thought of as a definition of a stable age composition since each age class is growing at the same rate. The relations developed in this chapter depend upon the assumption of a stable population.

9.2 STABLE POPULATION WITH INSTANTANEOUS RATES

The interrelations of population parameters for the case of instantaneous rates will be described first because the mathematics involved are simpler and more general. The symbols used have been defined in the first and second paragraphs of the preceding section.

The age composition at time t is $C_a = N_{at}/N.t.$ N_{at} is the number born a years ago times the probability of surviving to age a, which is l_a. By rearranging equation (9.1), the number born a years ago is $B_t e^{-ra}$. So $N_{at} = (B_t e^{-ra})l_a = bN._t e^{-ra} l_a$. Dividing both sides of this equation by $N._t$ gives the age composition

$$C_a = l_a b e^{-ra}. \tag{9.3}$$

The reproductive rate can be calculated from the following steps. The sum of the age composition over all ages equals 1, so from equation (9.3),

$$\int_0^\omega C_a \delta a = b\int_0^\omega l_a e^{-ra} \delta a = 1.$$

Solving this for b, the reproductive rate is

$$b = \frac{1}{\int_0^\omega l_a e^{-ra} \delta a}. \tag{9.4}$$

Also, since the number born is the total of those born to each age class at time t,

$$B_t = \int_0^\omega N_{at} b_a \delta a.$$

Dividing both sides by $N._t$ gives

$$b = \int_0^\omega C_a b_a \delta a. \tag{9.5}$$

Finally we come to a useful equation called the characteristic equation which may be derived by substituting C_a from equation (9.3) into (9.5) to give

$$b = \int_0^{\omega} l_a b e^{-ra} b_a \delta a,$$

and dividing both sides by *b*:

$$\int_0^{\omega} l_a b_a e^{-ra} \delta a = 1. \qquad (9.6)$$

The use of this equation is described in later sections.

9.3 AVERAGE RATES AND CONDITIONS IN A CONTINUALLY REPRODUCING POPULATION

Theory and Relations

The changes in a population in which births and deaths occur continually, as in human populations, are illustrated in figure 9.1. Instantaneous rates, as in the previous section, can be used for such populations, but in practice most rates are measured over a finite interval, such as a year, and are therefore average rates.

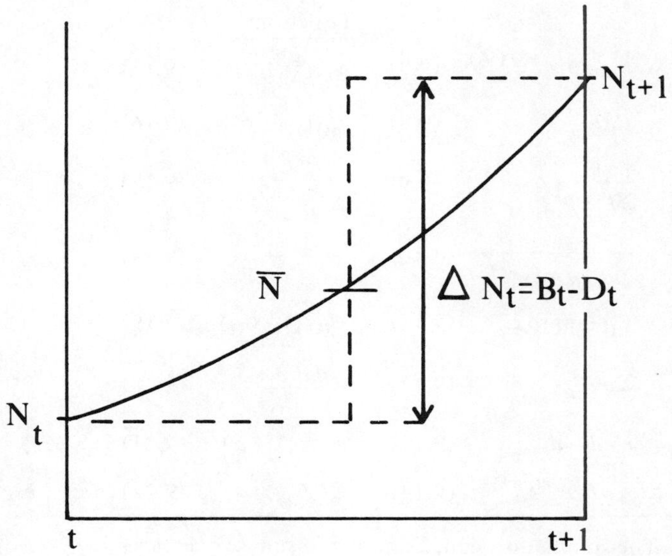

Figure 9.1. A population with continual reproduction and mortality. Symbols used in the text are illustrated.

For this section, and also for the following two, definitions and the equations showing the relations between the population characteristics are gathered into table 9.1. The equations are not proved, but for most of them the proofs are parallel to the derivations given in section 9.2, and the others can be obtained by manipulating the expressions given.

Table 9.1. Relations between Population Characteristics in Stable Populations

	Reproducing Continually		Reproducing in Short Seasons		Reproducing Once in a Lifetime
			Symbols and Definitions		
b	$B_t / \bar{N}_t$		$B_t / A._t$		B_t / A_t
d or s	$d = D_t / \bar{N}_t$		$s = A._{t+1} / N_t$		$s = A_{t+1} / B_t$
r	$\Delta N_t / \bar{N}_t$		-		-
g	-		$N._{t+1} / N._t$		B_{t+1} / B_t
C_a	$\bar{N}_{at} / \bar{N}._t$		$N_{at} / N._t$		1
			Equations		
$C_a =$	$l_{a+½} b\, e^{-r(a+½)}$	(9.7)	$l_a (\frac{b}{b+1}) g^{-a}$	(9.15)	-
$C_a =$	$C_0 l_a e^{-ra}$	(9.8)	$C_0 l_a\, g^{-a}$	(9.16)	-
$C_0 =$	$\frac{1}{\Sigma l_a e^{-ra}}$	(9.9)	$\frac{1}{\Sigma l_a g^{-a}}$	(9.17)	-
$b =$	$\frac{1}{\Sigma l_{a+½} e^{-r(a+½)}}$	(9.10)	$\frac{1}{\Sigma l_a g^{-a} - 1}$	(9.18)	$\frac{g}{s}$ (9.23)
$b =$	C_0 (approx.)	(9.11)	$C_0 / (1 - C_0)$	(9.19)	-
$b =$	$\Sigma b_a C_a$	(9.12)	$\frac{\Sigma b_a C_a}{1 - C_0}$	(9.20)	-
r or g:	$r = b - d$	(9.13)	$g = s(1 + b)$	(9.21)	$g = bs$ (9.24)
* $1 =$	$\Sigma l_{a+½} b_a e^{-r(a+½)}$	(9.14)	$\Sigma l_a b_a g^{-a}$	(9.22)	bsg^{-1} (9.25)

NOTE: The summation sign, Σ, means to sum over all ages.

*The Characteristic Equation.

The equations given for continually reproducing populations are approximations for the following reasons. The average number alive during a period is usually estimated by the number alive at the midpoint of the period. For example, the average number of age zero in year t, $\bar{N}_{0t}$, is estimated by the number less than one year old at time $(t + ½)$; these are the suvivors of those born between times $(t - ½)$ and $(t + ½)$. The Lexis diagram of figure 6.5 illustrates these time relations. The average survival time for this age class is approximately one-half year (it would equal one-half year if none died soon after birth). Similarly, for the $\bar{N}_{1t}$, the number between exact ages 1 and 2 at time $(t + ½)$, the approximate survival time is 1½ years. Therefore in the continually reproducing case, age is measured to the midpoint of each period, and this is indicated in some of the equations by the symbol $(a + ½)$. This is a better approximation than using age measured to the start of each period. However, in the derivation of other equations, such as (9.8), the l_a give a better approximation than do the $l_{a+½}$.

Applications

One of the simplest uses of the relations in a stable population is to compare the known age composition of a population with its expected age composition if it were stationary. The only information needed for the latter is the survivorship, the l_a or $l_{a+½}$, from life table studies. Equations (9.9) and (9.8) are used in that order, with $r = 0$, to calculate the expected structure of the hypothetical stationary population. Figure 9.2 compares the age composition calculated in this way with that of the actual population of Mexico revealing the effect of that country's high growth rate on its age structure.

Another illustration is given in example 9.1 where survival probabilities and a known growth rate are used to estimate the age composition and the birth and death rates of a population.

As was stated in the beginning of this chapter, the six parameters that describe a population are not independent of each other; knowing certain ones, the others can be calculated. Below are listed combinations of known characteristics from which the others can be calculated for the case of a continually reproducing stable population. The arrangement used here was that suggested by Keyfitz's (1968) treatment of the subject. When the life table is listed as one of the known characteristics, it is assumed that the $l_{a+½}$ to the midpoint of each age are known. If however only the l_a for the beginning of each age are known, which is often the case, reasonable approximations can be obtained by substituting these values, unless the periods measuring age are long in relation to the length of life. In the following expressions the summation sign, Σ, means to sum over all ages.

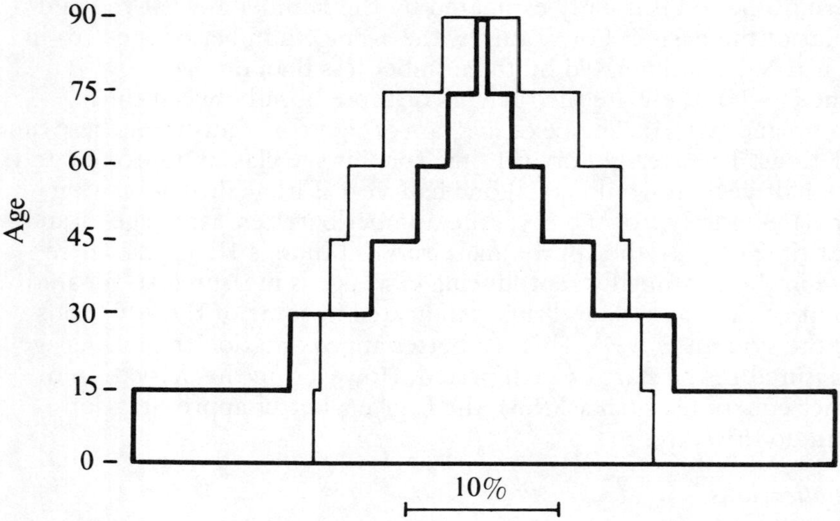

Figure 9.2. Age composition of Mexico (heavy line) compared with hypothetical composition calculated from survivorship rates (data from Keyfitz and Flieger 1971).

Ex. 9.1. Characteristics of a continually reproducing population. A hypothetical population has a survivorship curve like that of Figure 6.4b and continuous reproduction. Its annual growth rate is known to be $r = .04$. The survival probabilities are tabulated below for the midpoint of each age. The values of $e^{-r(a+½)}$ are also tabulated.

$a+½$	$l_{a+½}$	$e^{-r(a+½)}$	$a+½$	$l_{a+½}$	$e^{-r(a+½)}$
0.5	.92	.980	5.5	.58	.802
1.5	.78	.942	6.5	.46	.771
2.5	.74	.905	7.5	.28	.741
3.5	.70	.869	8.5	.13	.712
4.5	.64	.835	9.5	.04	.684

$$\Sigma l_{a+½} e^{-r(a+½)} = (.92 \times .980) + (.78 \times .942) + \ldots + (.04 \times .684) = 4.596$$

Using equation (9.10) to calculate the birth rate, $b = 1/4.596 = .218$. Using equation (9.7) to calculate C_a for ages 0, 1, 4, and 8:

$C_0 = .92 \times .218 \times .980 = .197;$

$C_1 = .78 \times .218 \times .942 = .160;$

$C_4 = .64 \times .218 \times .835 = .116;$

$C_8 = .13 \times .218 \times .712 = .020.$

From equation (9.13), $d = b - r = .218 - .040 = .178$

Examples 9.1 and 9.2 illustrate two of the following methods. Knowing *age composition* and the *life table*:

1. b can be estimated from equation (9.11);
2. d is approximated by $1 - \Sigma C_a l_{a+1\frac{1}{2}} / l_{a+\frac{1}{2}}$.
3. The results of (1) and (2) in equation (9.13) give r.

Knowing *age composition* and either r or d (as in example 9.2):

1. b can be estimated by equation (9.11);
2. The result of (1) with the known r or d in equation (9.13) gives the missing rate.
3. The age composition with r in equation (9.8) yields the set of l_a.

Ex. 9.2. Age composition and death rate known. In a continually reproducing, stable population the annual death rate is known to be 280/1000. The age composition in years is: $C_0 = .24$, $C_1 = .16$, $C_2 = .12$, $C_3 = .11$, $C_4 = .10$, $C_5 = .09$, $C_6 = .08$, $C_7 = .07$, $C_8 = .03$.
From equation (9.11) $b = .24$
From equation (9.13) $r = .24 - .28 = -.04$
From equation (9.8) $l_1 = C_1 / C_0 e^{-r} = .16 / .24 e^{.04} = .64$
$l_2 = .12 / .24 e^{.08} = .46$
$l_8 = .03 / .24 e^{.32} = .09$

Knowing *life table* $l_{a+\frac{1}{2}}$ and r (as in example 9.1):

1. b can be calculated from (9.10).
2. Result of (1) in equation (9.7) gives C_a.
3. Result of (1) in equation (9.13) gives d.

Knowing *life table* $l_{a+\frac{1}{2}}$ and either b or C_0:

1. b can be estimated from C_0, or vice versa, with equation (9.11).
2. In equation (9.9), for a given C_0 and set of l_a there is only one value of r that satisfies the equation, and it can be found by trial and error; section 9.8 describes an example of this type of calculation.
3. Results of (1) and (2) in equation (9.13) give d.
4. Results of (1) and (2) in equation (9.8) gives C_a.

Knowing *life table* $l_{a+\frac{1}{2}}$ and d:

1. C_0 of equation (9.9) can be replaced by its approximation from equation (9.11), b which equals $(d + r)$. As in the previous case, this can be solved for r by successive trials.
2. Results of (1) in $b = (d + r)$ gives b.
3. Results of (1) and (2) in equation (9.7) give C_a.

Knowing *life table* $l_{a+\frac{1}{2}}$ and the *age-specific reproductive rates* b_a:

1. From the characteristic equation (9.14) a value of r can be found by the method described in section 9.7.
2. Result of (1) in equation (9.10) gives b.

3. Results of (1) and (2) in equation (9.13) give d.
4. Results of (1) and (2) in equation (9.7) give C_a.
—or alternatively:
3. Results of (1) and (2) in equation (9.7) give C_a.
4. Result of (3) in equation (9.12) gives b.
5. Results of (1) and (4) in equation (9.13) give d.

9.4 AVERAGE RATES AND CONDITIONS IN A POPULATION WITH A SHORT REPRODUCTIVE SEASON AND WITH ADULTS REPRODUCING REPEATEDLY

Theory and Relations

Many vertebrates have a short reproductive period in one season of the year during which the population grows suddenly. From the end of one such season to the beginning of the next the population declines continually although not necessarily at a constant rate. In this section the mathematics will be simplified by assuming that all the births occur practically simultaneously, at the beginning of each year or period. Some individuals survive several periods and may reproduce several times. Figure 9.3 pictures the changes.

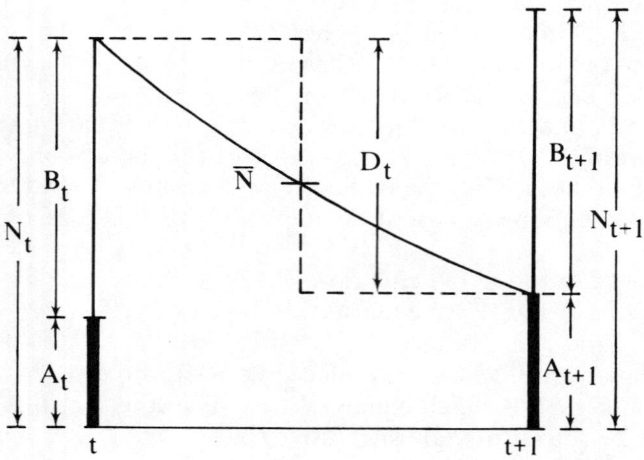

Figure 9.3. A population which reproduces only at the beginning of each period. Some individuals survive through several periods. Symbols used in the text are illustrated.

Some of the relations between population characteristics in this case are the same as in the previous one, but others are different for the following reasons: (1) births are concentrated at one time so the instantaneous growth rate varies; (2) age composition is measured at the time of reproduction and is not based on average numbers; (3) survival rate replaces the death rate and growth ratio replaces the growth rate in the interrelations. The equations describing the relations are in the center column of table 9.1.

Applications

The expected age composition of a stationary population may be calculated from the values of l_a by the same steps as in the previous case; that is, by using equation (9.17) and then (9.16) with g set equal to 1.

Example 9.3 is similar to the first example given in the preceding section and illustrates how known survivorship and growth rate are used to estimate the age composition and the birth and death rates of a population.

Ex. 9.3. Characteristics of a population with repeated, short reproductive seasons. Age is measured from the time of the reproductive season. The survival probabilities are tabulated below for the start of each age. The survivorship curve is similar to that used in example 9.1. The annual growth ratio is g = 1.041, which produces an increase equal to that of the previous example so that the two illustrations can be compared.

a	l_a	g^{-a}	a	l_a	g^{-a}
0	1.00	1.000	6	.54	.787
1	.84	.961	7	.36	.756
2	.76	.923	8	.20	.726
3	.72	.887	9	.08	.698
4	.67	.852	10	0	-
5	.61	.819			

$\Sigma l_a g^{-a} = 1 + (.84 \times .961) + \ldots + (.08 \times .698) = 5.116.$

Using equation (9.18) for the birth rate, $b = 1/(5.116 - 1) = .243$.

Using (9.15) for C_a for ages 0, 1, 4, and 8:

$b/(b + 1) = .243/1.243 = .196;$

$C_0 = 1 \times .196 \times 1 = .196;$

$C_1 = .84 \times .196 \times .961 = .158;$

$C_4 = .67 \times .196 \times .852 = .111;$

$C_8 = .20 \times .196 \times .726 = .029.$

From (9.21), $s = g/(1 + b) = 1.041/1.243 = .837$.

The various combinations of known parameters of a stable population used to calculate those that are unknown are described below.

Knowing the *age composition* and the *life table*:

1. b can be calculated from (9.19).
2. $s = \Sigma s_a C_a = \Sigma C_a l_{a+1} / l_a$.
3. Results of (1) and (2) in (9.21) give g.

Knowing *age composition* and either g or s:

1. b can be calculated from (9.19).
2. Result of (1) with known g or s in equation (9.21) gives the missing rate.
3. Previous results in equation (9.16) can give l_a.

Knowing *life table* l_a and g:

1. b can be calculated from equation (9.18).
2. Result of (1) in equation (9.21) gives s.
3. Result of (1) in equation (9.15) yields C_a.

Knowing *life table* l_a and b (as in example 9.4):

1. In equation (9.18) for a given set of b and l_a only one value of g will satisfy the equation. This can be determined by successive approximations (see next section for a related problem).
2. Result of (1) in equation (9.21) will yield s.
3. Result of (1) in equation (9.15) will yield C_a.

Ex. 9.4. Stable population with known survivorship and birth rate. A population of animals which reproduce in late spring was studied over several years during which the proportions of each age stayed constant. Age-specific survival rates were determined, yielding the following l_a: $l_0 = 1$; $l_1 = .58$; $l_2 = .50$; $l_3 = .44$; $l_4 = .38$; $l_5 = .31$; and $l_6 = .20$. The general birth rate was estimated to be $b = .51$.

A trial value of $g = 1$ (i.e., for a stationary population) gives a sum $\Sigma l_a 1^{-a} = 1 + .58 + \ldots + .20 = 3.41$.

From equation (9.18), $1/(3.41 - 1) = .415$, which is less than $b = .510$.

A second trial value of $g = 1.1$ is assumed. Using this, $\Sigma l_a 1.1^{-a} = 1 + .58/1.1 + .50/1.1^2 + \ldots + .20/1.1^6 = 2.835$. This sum in equation (9.18) gives .545 which is greater than b. By interpolation, a third value of $g = 1.07$ is selected, and, by the same steps as before, this gives the number .505 which is satisfactorily close to $b = .51$.

From equation (9.21), $s = 1.07/1.51 = .71$.

Using equations (9.15) and (9.16): $C_0 = .51/1.51 = .34$;
$C_1 = .34 \times .58 \times 1.07 = .21$, and so on.

Knowing *life table* l_a and s:

1. From equation (9.21), b of equation (9.18) can be replaced with $(g/s - 1)$, and as in the previous case only one value of g, found by trial, will satisfy the equation.

2. The result of (1) in equation (9.21) gives b.
3. Results of (1) and (2) in equation (9.15) will yield C_a.

Knowing l_a and b_a:

1. The characteristic equation (9.22) can be solved for g by the method described in the next section.
2. The result of (1) in equation (9.18) gives b.
3. The results of (1) and (2) in equation (9.15) yield C_a.
4. Results of (1) and (2) in equation (9.21) gives s.

Another application of the equations of this section is their use to assess estimated survival rates; the idea explained here applies also to other sections, but an example is included now because many of the species for which survival rates have been calculated have the reproductive pattern treated in this section. The core of the method is to test if equation (9.22) is satisfied by substituting the estimated values of l_a calculated from the successive survival rates and information on the birth and growth rates of the population. For example, if a population has survival s_0 to age 1 when first breeding occurs, a constant s_A thereafter, and also a constant birth rate, and if the population is believed to be stationary, then equation (9.22) becomes

$$1 = b\Sigma l_a = b(s_0 + s_0 s_A + s_0 s_A^2 + s_0 s_A^3 + \ldots) = bs_0/(1 - s_A).$$

If the known or estimated values of b, s_0, and s_A are reasonable, they should approximately satisfy this equation. A more realistic example is given in example 9.5.

Ex. 9.5. Testing the plausibility of estimated survival rates. Survival rates have been estimated for a population from data on the recovery of marked individuals. The estimated rates produce the expected survival probabilities, l_a, shown below. Studies of reproduction show that the age-specific reproductive rates are as tabulated under b_a. The population is thought to be stationary, so $g = 1$ and $g^{-a} = 1$. The purpose of the following calculations is to see if equation (9.22) is satisfied by the estimates and assumptions given above.

a	l_a	b_a	$l_a b_a g^{-a}$
0	1.00	0	0
1	.24	2	.48
2	.12	3	.36
3	.08	3	.24
4	.06	2	.12
5	.04	0	0
6	0	-	-
			1.20

The $\Sigma l_a b_a g^{-a} > 1$, so either the estimated survival rates are too high, or the population is not stationary but is increasing (i.e., $g > 1$).

Applications in Age Groups of Unequal Duration

In many species the individuals can be classified only into age groups some of which span more years than others; an example is the herring gull population described by Kadlec and Drury (1968), in which the recognizable classes are juveniles (in their first year), intermediates (in their second or third year), and adults (fourth year and older). The general case can be explained conveniently by assuming that there are two recognizable groups: intermediates, I (which include n years, the oldest being designated the I_n) and the adults, A, which are older. The population is assumed to be stable, with a growth ratio of g, and a constant survival rate s for both intermediates and adults. Then $g = I._{t+1}/I._t = A_{t+1}/A_t$, and $A_{t+1} = s(A_t + I_{nt})$. Beginning with these relations it can be shown that

$$\frac{I._t}{A_t} = \left(\frac{g}{s}\right)^n - 1. \tag{9.26}$$

The use of this equation is illustrated in example 9.6.

Ex. 9.6. Application to a population with unequal age classes. Kadlec and Drury (1968) observed that the intermediate age (second and third year) herring gulls composed 17 percent, and the adults 68 percent of the population. They estimated that the population had been doubling every 12 to 15 years. Using a doubling time of 13 years in equation (8.6), the growth ratio $g = e^{(.693/13)} = 1.055$. The survival rate can then be calculated from equation (9.26) with $n = 2$:

$$.17/.68 = (1.055/s)^2 - 1$$

$$s = .94.$$

9.5 AVERAGE RATES AND CONDITIONS IN A POPULATION WITH REPRODUCTION ONLY ONCE IN A LIFETIME

Theory and Relations

Many insects illustrate this third type of reproductive pattern in which adults mature, reproduce, and then die, usually at a particular season of the year. Figure 9.4 shows this pattern. For all practical purposes related to the study of such a population, the adults may be considered to die immediately after reproducing, so that only one age class effectively is present. Because of this, the relations in this case are simpler than in the preceding two. The useful equations are easily derived from those in the preceding section. The three shown in the right column of table 9.1 are versions of just one equation.

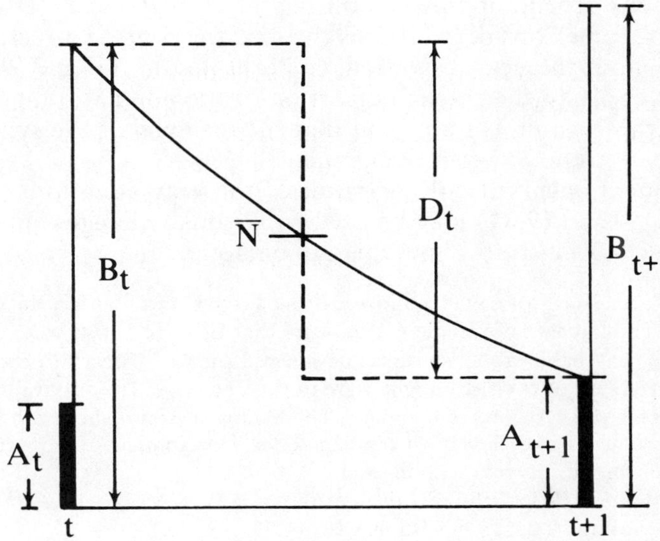

Figure 9.4. A population in which adults reproduce once and then die. Symbols used in the text are illustrated

Applications

It is obvious that if any two parameters are known, the third can be calculated.

9.6 RECRUITMENT AND EXPLOITATION IN STABLE POPULATIONS

Recruitment is discussed in chapter 7, section 7.3, where the recruitment rate is defined as the number added to the catchable stock of a species in proportion to the total size of that stock. That section contains the statement that the birth rate may be considered a special case of the recruitment rate. The reverse is partially true. If the recruits are designated as being of age 0—that is, the youngest age class in the catchable stock—then the recruitment rate can be defined just as is the birth rate in the first line of table 9.1, and the recruitment rate will have many of the same relations within the catchable portion of the population as does the birth rate in the whole population. These will be explained not by changing the symbols, but by defining those in table 9.1 so that they apply, in this section, to the catchable stock. Thus b becomes the recruitment rate, B the number of new recruits, N_a the number of age a counting from age 0 for the recruits, and so on.

The relations depend upon when during the annual cycle recruitment occurs; to be considered first is the case for continual recruitment throughout the year. Equations (9.7) through (9.11) and (9.13) of table 9.1 apply to the catchable portion of the population, providing that this portion is stable and that the meaning of the symbols is as described in the previous paragraph.

If recruitment occurs at only one time of the year, equations (9.15) through (9.19) and (9.21) may be used with similar changes in meaning. Example 9.7 illustrates how this can be done.

Ex. 9.7. Information from age composition and survival rates within the catchable stock. The figures in example 6.4 indicate that fish of age five were completely in the catchable stock, so this is designated the age of recruitment. Fifty-three percent of the fish caught were in this age class. The survival rate estimated from the catch curve is $s = .41$. The fishing season is short and follows summer, when growth is most rapid, so equations applicable to recruitment at one time of the year will be used.

From equation (9.19), the recruitment rate = $.53/(1 - .53) = 1.13$.

From equation (9.21), $g = .41(1 + 1.13) = .87$.

The characteristics of a stable population are important to a fisheries or wildlife manager because he is often trying to control mortality by catching or hunting so as to obtain the maximum sustained yield from some population. If successful, he has created a stationary population in which the harvest is the major cause of death. The conditions in the catchable stock can then be estimated by setting r equal to zero, or g equal to 1, in the equations of table 9.1 that apply to recruitment and the catchable stock. For example, the proportion of each age class in the stock is directly proportional to the probability of surviving to that age (equations [9.7] and [9.8], or [9.15] and [9.16]). If recruitment is continual throughout the year, the recruitment rate equals the death rate from all causes (9.13) and is approximately equal to the proportion of recruits in the stock (9.11). If recruitment occurs at one season of the year, the mortality rate from all causes, related to the recruitment and the proportion of recruits, is $m = b/(b + 1) = C_0$ (from equations [9.21] and [9.19]).

These equations are of little use, however, for deciding on optimum fishing or kill rates for a number of reasons. One is that they apply only to the catchable stock, and the recruitment rate is largely determined by the whole population's birth rate and the survival from birth to recruitment. A second is that a change in mortality will change the age composition of the reproductive ages which will affect the general birth rate (equations [9.12] and [9.20]). A third reason is that these relations do not take into account any density-dependent effects on either survival or reproduction from changes in the size of the population. More effective methods for predicting the optimum yield are outlined in the next chapter.

9.7 GENERATION TIME

Generation time is easy to define for species which reproduce just once in their life cycle; it is the time from birth to reproduction. But for species which reproduce repeatedly, with surviving offspring of different ages and with offspring often overlapping their parents in reproduction, the generation time is more difficult to define and measure. Three different definitions are given below.

The first arises from the net reproductive rate, R_0, in equation (7.5). Since this is the ratio of female births of offspring to the same of their mothers, by using equation (9.1) it can be written $R_0 = e^{rT}$, where T is a measure of generation time:

$$T = \frac{\ln R_0}{r} \tag{9.27}$$

This concept has two weaknesses: it is meaningless for a stationary population when $r = 0$, and it has no biological basis.

A second measure is the *cohort generation time*, defined as the mean age of the females of a cohort at the birth of their daughters. The cohort generation time is

$$T_c = \frac{\Sigma_a \, a l_a b_a}{\Sigma_a \, l_a b_a}, \tag{9.28}$$

the sum of the mothers' ages at all female births divided by the number of all female births.

Ex. 9.8. Calculation of the cohort generation time. The figures in example 9.5 will be used by assuming that the l_a column applies to a female cohort and the b_a is the rate for births of daughters.

$$\Sigma a l_a b_a = (1 \times .48) + (2 \times .36) + (3 \times .24) + (4 \times .12) = 2.40 \text{ years}$$

$$\Sigma l_a b_a = 1.20$$

$$T_c = \frac{2.40 \text{ years}}{1.20} = 2 \text{ years.} \qquad \text{(eq. 9.28)}$$

The third measure is similar, being the mean age of mothers in a population (not a cohort) at the birth of their daughters. The *mean length of a generation* (Leslie 1966) is

$$\overline{T} = \frac{\Sigma_a aC_a b_a}{\Sigma_a C_a b_a}\,.$$

This equation can be reduced by using equations (9.3) and (9.6) to

$$\overline{T} = \Sigma_a a l_a b_a e^{-ra}. \qquad (9.29)$$

In a stationary population $T_c = \overline{T}$.

Generation time has had little application to ecological problems.

Ex. 9.9. Calculation of the mean length of a generation. Using again the figures in example 9.5 and, in addition, the value of $g = 1.1$*, the mean length of a generation can be calculated from equation (9.29).

$$\overline{T} = 0 + (1 \times .24 \times 2 \times 1.1^{-1}) + (2 \times .12 \times 3 \times 1.1^{-2}) +$$
$$(3 \times .08 \times 3 \times 1.1^{-3}) + (4 \times .06 \times 2 \times 1.1^{-4}) = 1.9 \text{ years}$$

*This value approximately satisfies equation (9.21).

9.8 ESTIMATION OF r IN A STABLE POPULATION

Survivorship probabilities from a life table and age-specific reproductive rates determine the growth rate of a population providing that it has attained a stable age distribution. The growth rate, r, can be calculated from the characteristic equation, which has been given in various forms, equations (9.6), (9.14) and (9.22). The usual form for estimating r is

$$1 = \Sigma_{a=0}^{\omega} l_a b_a e^{-ra}\,, \qquad (9.30)$$

$$= l_0 b_0 + l_1 b_1 e^{-r} + l_2 b_2 e^{-2r} + \ldots l_\omega b_\omega e^{-\omega r}.$$

Usually the survivorship, l_a, is that of females only, and the reproductive rate, b_a, is that for the birth of daughters to mothers of age a. This will be understood in the discussion that follows.

An intuitive understanding of equation (9.30) can be obtained by first considering the situation when $r = 0$, when the characteristic equation reduces to a definition of the net reproductive rate (equation 7.5), $R_0 = \Sigma\, l_a b_a = 1$ when growth is zero. If $r > 0$, $\Sigma\, l_a b_a > 1$; but as r increases each value of the exponential function e^{-ra} decreases so as to keep the sum, $\Sigma\, l_a b_a e^{-ra}$, equal to 1. If $r < 0$, the balancing works in the opposite direction.

In species with short reproductive seasons, when numbers increase in spurts separated by periods of decline, population growth is measured by the growth ratio, g (equation 8.5), more reasonably than by the rate r. But for reasons to be given later there are advantages to working with the characteristic equation in the form of (9.30), and this can be done easily by the transformation $g = e^r$.

The characteristic equation (9.30) cannot be solved directly for r. Some method of trial and error or successive approximation must be used to find a value of r which satisfies the equation (i.e., which makes the sum equal 1). Keyfitz (1968, pp. 108–14) describes a few methods for numerical solution of the equation. The usual method used in the ecological literature and described by Birch (1948) can be explained graphically by figure 9.5. The sum $\Sigma l_a b_a e^{-ra}$, is treated as a dependent variable, a function of the independent variable r. A trial

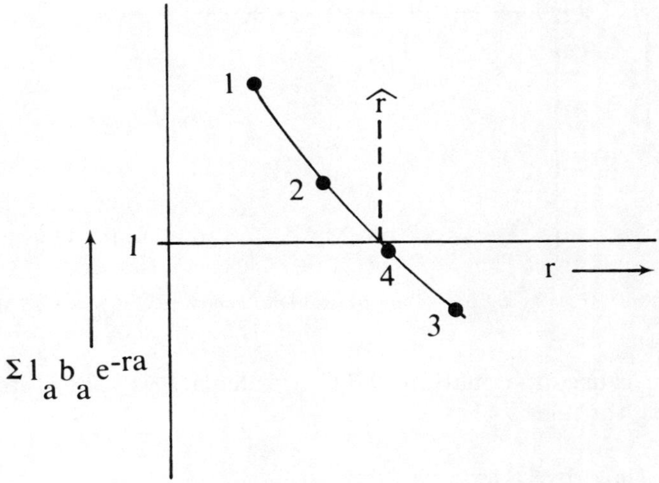

Figure 9.5. Values of the sum of equation (9.30) plotted as a function of r. The dots on the curved line represent trial values of r, the numbers indicating the sequence in which they might have been tested.

value of r is used to calculate the corresponding sum, and then by successive approximations and perhaps by interpolation, a value of $\hat{r}$ is found which makes the sum approximately equal to 1. The decision when to stop the successive approximations depends upon the desired number of significant figures in $\hat{r}$.

The estimation can be speeded by transforming equation (9.30) by natural logarithms,

$$y = \ln \Sigma l_a b_a e^{-ra} = 0 . \qquad (9.31)$$

This changes the curved line of figure 9.5 to an almost perfectly straight line, which allows efficient use of Newton's Method of approximation (described in standard calculus texts). The method can be outlined with the aid of figure 9.6. A trial value of r is selected

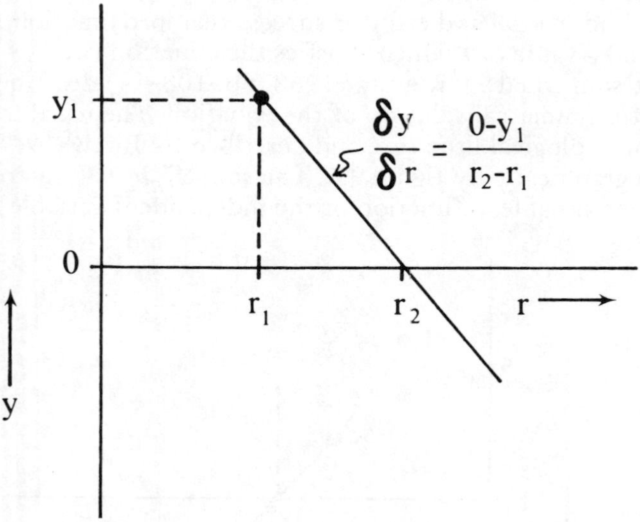

Figure 9.6. Graph of $y = ln \ \Sigma \ l_a b_a e^{-ra}$ as a function of r and the use of Newton's Method to approximate $\hat{r}$.

and this r_1 is used in equation (9.31) to calculate y_1. The slope of the line passing through y_1 is

$$\frac{\delta y}{\delta r} = \frac{\delta}{\delta r}(\ln \Sigma l_a b_a e^{-ra}) = \frac{-\Sigma a l_a b_a e^{-ra}}{\Sigma l_a b_a e^{-ra}}, \qquad (9.32)$$

evaluated for $r = r_1$. This line projected to $y = 0$ determines r_2, which can be estimated because $\delta y / \delta r = (0 - y_1)/(r_2 - r_1)$ and $\delta y / \delta r$ is known from (9.32). The process can be repeated, using r_2 to calculate y_2, evaluating (9.32), and estimating r_3, and so on. Because y is almost a linear function of r, few iterations are needed to obtain an estimate of $\hat{r}$ accurate to four or five significant figures.

Estimation of $\hat{r}$ begins by somehow selecting a first trial value, r_1. There is a convenient way to do this by using the same data as are required for the characteristic equation. From a, l_a, and b_a the cohort generation time, T_c, can be calculated from equation (9.28). The denominator of equation (9.28) is R_0 (see equation [7.5]). Substituting T_c for T in (9.27) and solving for $r = ln \ R_0/T_c$ gives an approximation which can be used for r_1. This will probably be lower than the true r, and Laughlin suggests (1965) that it should be multiplied by 1.155 to give a better approximation.

The procedure for estimating r from the characteristic equation can be organized as follows. The steps are placed in a logical order; a person familiar with computers or programmable calculators may change the order for more efficient computation.

Table 9.2. Calculations for the Estimation of r from the Characteristic Equation. (See text for explanation.)

Step 1			Step 2		Step 4	Step 5	
1	2	3	4	5	6	7	8
a	l_a	b_a	$l_a b_a$	$a l_a b_a$	e^{-ra}	$l_a b_a e^{-ra}$	$a l_a b_a e^{-ra}$
0	1	0	0	0	1	0	0
1	l_1	b_1	$l_1 b_1$	$1 l_1 b_1$	e^{-r}	$l_1 b_1 e^{-r}$	$1 l_1 b_1 e^{-r}$
2	l_2	b_2	$l_2 b_2$	$2 l_2 b_2$	e^{-2r}	$l_2 b_2 e^{-2r}$	$2 l_2 b_2 e^{-2r}$
.	.	.	.	.	.	.	.
.	.	.	.	.	.	.	.
ω	l_ω	b_ω	$l_\omega b_\omega$	$\omega l_\omega b_\omega$	$e^{-\omega r}$	$l_\omega b_\omega e^{-\omega r}$	$\omega l_\omega b_\omega e^{-\omega r}$
			$\Sigma l_a b_a$	$\Sigma a l_a b_a$		$\Sigma l_a b_a e^{-ra}$	$\Sigma a l_a b_a e^{-ra}$

1. Prepare a table as illustrated in table 9.2. Complete the first three columns, which are the data for the age, (female) survivorship, and age-specific (female) birth rate.
2. Calculate each product in column 4 and column 5 and sum each column.
3. For the first trial estimate of r, calculate:
   A. $ln\ R_0 = ln \Sigma\, l_a b_a$ (this sum is the sum of column 4);
   B. $T_c = \Sigma\, a l_a b_a / \Sigma\, l_a b_a$ (sum of column 5 divided by sum of column 4);
   C. For first trial, $r_1 = ln\, R_0 / T_c$, or use 1.155 times this figure.
4. Using $r = r_1$, calculate the values in column 6 (with a table of exponential functions).
5. Calculate each product in column 7 and column 8 and sum each column. The sum of column 7 should be in the neighborhood of 1.
6. From a table of natural logarithms determine the logarithm of the sum of column 7; this is the y of equation (9.31).
7. Using r_1 (step 3) and y_1 (step 6), calculate r_2 from the following

(obtained by equating (9.32) to $(0 - y_1)/(r_2 - r_1)$; see figure 9.6):

$$r_2 = r_1 + \frac{y_1 \Sigma l_a b_a e^{-ra}}{\Sigma a l_a b_a e^{-ra}} . \qquad (9.33)$$

8.–11. With $r = r_2$, repeat steps (4) through (7) to obtain a better approximation r_3. Continue until successive approximations do not change the estimated value of r in the number of significant figures desired.

For most wild populations of animals, estimation of r to an accuracy of more than three or even two significant figures is rarely justifiable, owing to the lack of accurate knowledge about most life history parameters. Reasoning strictly, the b_a and the l_a must come from the same population, and the l_a from a current life table (chap. 6, sec. 6.6). Many estimates of r have been based on survival rates obtained from the age composition of assumed stationary populations (Caughley and Birch 1971), and unless there is evidence that these age-specific survival rates also hold in non-stationary populations, the results are of course questionable.

9.9 THE INTRINSIC RATE OF POPULATION GROWTH

A useful concept in ecology and evolution is that each species has a potential rate of increase that is determined by genetically controlled characteristics such as fecundity, age of maturation, and longevity (see Cole 1954 for a review of the subject). A good term for this concept is the "intrinsic population growth rate," and a practical definition is that it is the maximum possible growth rate of a population with a stable age distribution.

The adjective "intrinsic" implies that this rate is a characteristic of a species (or a deme, an interbreeding population). The words "maximum possible" mean that the rate can be attained only under one set of conditions and when no resources are limiting, which could in this respect therefore be called the optimum environment. The condition that the population have a stable age distribution is necessary because otherwise the growth rate would depend very much on the immediate age structure which would itself be changing due to the growth process. The above definition and interpretation of its meaning are consistent with those generally appearing in the ecological literature.

The most frequently used symbol for the intrinsic population growth rate is r_m.

Estimation of the intrinsic growth rate is done by steps described in the previous section. The age-specific reproductive rates, b_a, and

Ex. 9.10. Estimation of r from optimum survivorship and age-specific fertility. The data are for the snowshoe hare (Tanner 1975), modified slightly to make a better example. The steps are numbered as in the previous description.

Step 1			Step 2		Step 4	Step 5	
1	2	3	4	5	6	7	8
a	l_a	b_a	$l_a b_a$	$a l_a b_a$	e^{-ra}	$l_a b_a e^{-ra}$	$a l_a b_a e^{-ra}$
1	.20	6.0	1.200	1.200	.63251	.75901	.75901
2	.09	6.4	0.576	1.152	.40007	.23044	.46088
3	.04	6.4	0.256	0.768	.25305	.06478	.19434
4	.02	6.2	0.124	0.496	.16005	.01985	.07940
5	0	-	0	0	-	0	0
			2.156	3.616		1.07408	1.49363

Step 3: A. $\ln R_0 = \ln 2.156 = .76825$

B. $T_c = \dfrac{3.616}{2.156} = 1.67718$

C. $r_1 = \dfrac{.76825}{1.67718} = .45806$

Step 6: $\ln 1.07408 = .07146$

Step 7: $r_2 = .45806 + .07146 \dfrac{1.07408}{1.49363} = .50945$

Steps 8–11, with $r_2 = .50945$:

a	e^{-ra}	$l_a b_a e^{-ra}$	$a l_a b_a e^{-ra}$
1	.60083	.72100	.72100
2	.36099	.20793	.41586
3	.21689	.05552	.16656
4	.13032	.01616	.06464
5	-	0	0
		1.00061	1.36806

Step 10: $\ln 1.00061 = .00061$

Step 11: $r_3 = .50945 + .00061 \dfrac{1.00061}{1.36806} = .50990$

The important points of the above calculations are that the first approximation of $r = .45806$ yields a $\Sigma l_a b_a e^{-ra} = 1.07408$, and the second approximation of $r = .50945$ improves this sum to 1.00061. The third approximation of $r = .50990$ yields a sum of 1.0000001 (calculations not shown). The original data are not accurate enough to warrant a result of more than two significant figures (i.e., $r = .51$), but this method proceeds faster if several significant figures are retained until the final result.

the survivorship rates, l_a, used in this calculation are usually those believed to be the highest possible for the species; that is, those that would be attained in an optimum, unlimited environment. It is possible, of course, that the conditions producing the highest reproductive rates would not result in the best survival, and vice versa, and combining the two highest sets of values would give a falsely high estimate of the intrinsic rate. The only procedure in such a case would be to combine b_a and l_a from the same environmental conditions and find r_m by trial and error with different combinations.

A growing body of literature discusses the effects of separate life history or population characteristics on the intrinsic growth rate. Cole's (1954) paper was one of the first and provides a review to that date. Goodman (1971), Keyfitz (1971), and Coale (1972) present mathematically oriented analyses of the question, the latter two concentrating on human populations. Giles (1969) and Myers (1971) include diagrams to reveal the interactions of a few variables on the intrinsic rate. Gadgil and Bossert (1970) consider the problem from an evolutionary viewpoint.

10 Prediction of Population Changes

10.1 PURPOSES OF POPULATION MODELS

Models are used in ecology to represent or to imitate complex processes and simultaneously to reduce the complexities to simple and more manageable procedures. A population model is intended to behave quantitatively like the numbers of a population, and the model is constructed either by inferring the processes from the study of populations or by making reasonable assumptions. These are incorporated into a set of equations or a series of prescribed calculations. Thus the model becomes an abstraction of population processes. If the model behaves as does a population, the assumptions and conditions included must somehow parallel the processes occurring in populations. And conversely, if a model is a good imitation, it will correctly predict changes in a real population.

Population models can then be seen to have two major functions. One is to abstract and idealize population processes for better understanding. A model built for this purpose may be quite simple. It is used as a hypothesis; the predictions or deductions based on the model are compared with "the real thing." If they do not fit, something is wrong with the structure or conditions of the model. If they do, the model can be either simplified or elaborated to find the limits of the agreement and thus to get better insights into the behavior of populations. Models designed with this goal are referred to by May (1973) as "general" or "strategic" models. He compares their role in ecology with that of the perfect crystal in physics; the latter does not exist but the concept helps to direct thinking about crystals.

A second goal of models is to predict changes in specific populations for management or similar purposes. A model for this function needs to include as many factors as are known in order to make an accurate prediction. This type of model therefore tends to be complex (of course the most complex and realistic model is the population itself, which, however, cannot predict itself). Complex models with many interactions are the subject of "systems analysis." Watt (1968) describes aspects of such modeling applied to ecology and resource management, and Conway (1973) surveys the use of models in insect pest control and also comments on the nature and general purpose of models.

More general treatment of population models may be found in Maynard Smith (1974), Poole (1974), and in a symposium volume edited by Bartlett and Hiorns (1973), another by Jeffers (1972), and a third by May (1976). The citations mentioned in this introduction are only a small sampling of the literature on population models used in ecology; several hundred citations on population growth, modeling, and simulation are listed in a bibliography assembled by Schultz and others (1976). This chapter is not intended to be a survey of this growing field. It aims only to introduce the subject by outlining four major kinds of models: differential equations, difference equations, matrix models, stochastic models. It is limited to populations of single species. Models that imitate the behavior of two interacting species, such as a prey and its predator, are another story; May's (1974) book on interacting species and ecosystem stability is a good introduction to that subject, and several chapters of the collection edited by him (May 1976) are about aspects of this subject.

10.2 MODELS USING DIFFERENTIAL AND DIFFERENCE EQUATIONS

The simplest model of population growth assumes that the growth rate r is constant. Expressed as a differential equation and its solution, it is $dN/dt = rN$, from which $N_t = N_0e^{rt}$, where N_t is the number at any time t and N_0 is the number when $t = 0$. This equation predicting exponential growth is not realistic, because it would require unlimited resources to allow a constant r, but it may apply to the special case of a very small population in a benign environment.

Many populations have shown a growth which was at first nearly exponential when numbers were small, which later gradually slowed and finally stopped when the number attained an upper limit, producing an S-shaped growth curve (examples of animal populations showing this form of growth may be found in Allee *et al.* 1949). A model widely known as the "logistic" produces such a curve. The differential equation for the logistic model is

$$\frac{\delta N}{\delta t} = r_m N\left(\frac{K - N}{K}\right). \tag{10.1}$$

There are two parameters: r_m, the intrinsic population growth rate; and K, the carrying capacity (i.e., the maximum population that can be maintained by the particular environment). The operation of the model is illustrated by figure 10.1. When N is very small, the quantity

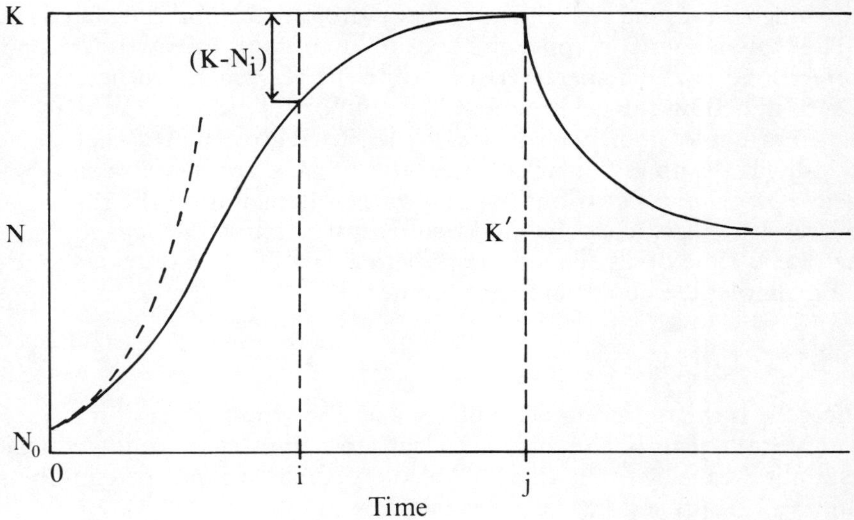

Figure 10.1. The logistic model of population growth (equations 10.1 and 10.2). The dashed curve extending upward from N_0 illustrates exponential growth at a constant growth rate. At time j the carrying capacity is suddenly decreased to K'.

in parenthesis in equation (10.1) is approximately 1, and the population is growing almost exponentially with a per capita growth rate equal to r_m. As N increases, the quantity in parenthesis decreases and so does the population growth rate. At time i in the figure the proportion of unused resources or still empty habitat will be $(K - N_i)/K$. This suggests a useful way to look at the logistic model: the population growth rate is proportional to the fraction of the resources not used at that instant. When N equals K, growth is zero; the numbers actually approach K asymptotically in time. Figure 10.1 also illustrates the effect of a sudden decrease in the carrying capacity, from K to K', and the effect of this on N. The logistic model predicts decreases as well as increases.

Equation (10.1) can be integrated to give

$$N_t = \frac{K}{1 + ae^{-r_m t}},$$

where

$$a = \frac{K - N_0}{N_0}. \tag{10.2}$$

Knowing r_m, K, and N_0, the number at any time can be calculated.

The logistic model is appealing because it contains just two parameters: one, r_m, is a characteristic of the species, and the other, K, is determined by the environment. One biological weakness is that any change in N immediately affects the actual growth rate; that is, all individuals, no matter what their sizes or ages, are the same in their reproductive contributions and in their demands on the resources. The logistic can be modified by introducing time lags so that the rate at time t will depend upon the number at some previous time. For example, the equation may become

$$\frac{\delta N}{\delta t} = r_m N_{t-T_1} \left(\frac{K - N_{t-T_2}}{K}\right), \tag{10.3}$$

where T_1 is the reproductive time lag and T_2 is the time lag for exploitation of resources. The effect of most time lags is to introduce instability, causing oscillations of various degrees around the equilibrium K. This process has been reviewed by May *et al.* (1974).

Other modifications of the logistic which have been useful or suggestive have been (1) replacing N with a measure of the population's biomass (Frank 1960, Gallopin 1971); (2) incorporating the means and variances of age and mass (Streifer and Istock 1973); (3) replacing the effects of population density with the rate of food utilization (Smith 1963); and (4) introducing a threshold below which growth is exponential and above which the slowing of the growth rate depends on the excess of numbers above the threshold (Wiegert 1974). Although the logistic model has been frequently criticized for its simplistic assumptions, it continues to be useful in modeling populations.

The logistic model suggests two ways of estimating the intrinsic population growth rate of a species. One is to measure the growth of a population so small that the observed growth is not reduced significantly by density and the population is growing almost exponentially. The other depends upon knowledge of K, the carrying capacity. Caughley and Birch (1971) suggest that this might be obtained from a population which has been stationary for some time at a level taken to be K, and then if the population can somehow be reduced sharply and allowed to recover, its growth during the recovery can supply data for estimating r_m. For doing this, equation (10.2) can be modified to $ln(K/N_t - 1) = ln\ a - r_m t$. This is in the form of a linear regression, with t being the independent variable and r_m the regression coefficient, which can be estimated by the usual methods. Caughley and Birch (1971) give a simple example.

Differential equations describe changes which are continuous, and are therefore best applied to populations like those of man, wherein reproduction and death occur at all seasons and generations overlap. They do not fit the pattern of many insect species, where the adults die soon after reproducing and one generation replaces another. For this situation, difference equations are appropriate. A difference equation can be generalized as $N_{t+1} = f(N_t)$ where N and t are integers. The difference equation which is parallel to the differential equation for exponential growth is $N_{t+1} = gN_t$, where g is the growth ratio of equation (8.5). The prediction of population change is then accomplished by iterative use of this equation, which would of course produce a geometric series. Most differential equations can be converted into difference equations. For example, the comparable difference equation for the logistic model is

$$N_{t+1} = N_t\left(\frac{Ke^{r_m}}{K + N_t(e^{r_m} - 1)}\right). \tag{10.4}$$

The behavior of a difference equation model cannot always be predicted intuitively. Depending upon the relative size of the growth rate, the model may have a stable equilibrium, a stable oscillation among two or more points, or it may be "chaotic." As the growth rate increases, instability increases. This subject is reviewed by May (1976), who cites the original papers and points out parallels in natural populations.

Another type of difference equation is $N_{t+1} = N_t + \Delta N_t$, where ΔN_t is a function of N_t and Δt. This is equivalent to approximating $\delta N/\delta t$ by $\Delta N/\Delta t$, and is a useful way to approximate the results of a differential equation which cannot be solved by integration. An illustration of the procedure is in example 10.1. This kind of difference equation must be used cautiously in predicting population changes, because it assumes that the growth over the period Δt is determined only by N_t, the number at the beginning of the interval as well as other conditions existing then. The effect is very much like that of introducing time lags into a differential model: stability is decreased. The inaccuracies can be reduced by making Δt as small as is practical;

the limitations here are usually those of time and energy for making the repeated calculations. Computers almost remove these limitations and are ideal instruments for predictions using difference equations; they can produce growth curves appearing as smooth as the plot of a differential equation.

Ex. 10.1. Application of a difference equation to the logistic model. Equation (10.1) is rewritten to read: for $\Delta t = 1$, $\Delta N_t = r_m N_t (K - N_t)/K$. Letting K = 1000, $r_m = .1$, $\Delta N_t = .1N_t(1 - .001N_t)$. Starting with $N_0 = 50$ and rounding the results to the nearest tenth, the results are partially tabulated below:

t	N_t	ΔN_t	$N_t + \Delta N_t$
0	50	4.8	54.8
1	54.8	5.2	60.0
2	60.0	5.6	65.6
.	.	.	.
.	.	.	.
72	986.9	1.3	988.2
73	988.2	1.2	989.4
74	989.4	1.0	990.4

Difference equations can be used to construct "stock-recruitment" or reproductive curves proposed and developed by Ricker (1954, 1975). One is illustrated in figure 10.2. On the horizontal axis is N_t, the stock or number of reproducing adults at time t, and on the vertical axis is the recruitment or the number one generation later, N_{t+1}. The straight line with a slope of 1 is the locus of all equilibrium points where the recruits exactly replace the previous number. All points above and to the left of this line are for growing populations, and all those on the opposite side are for declining populations. The curve of figure 10.2 is that produced by the logistic model, and the intersection of the two lines is at K. The diagram is used to predict population growth by reading from N_t up to the curve and then left to find N_{t+1}; this becomes the next N_t (population or stock), and the process is repeated.

An explanation of stock-recruitment models and the development of the curves used in fishery studies may be found in Paulik and Greenough (1966). Watt (1968, chaps. 3–5) illustrates several uses of these models.

10.3 MODELS USING AGE-DEPENDENT REPRODUCTIVE AND SURVIVAL RATES

If the parameters of specific survival and reproductive rates are known, and the numbers of each age in the population, these can be used to predict the structure and size of the population at some future time. The number of each age times the survival rate for that age produces

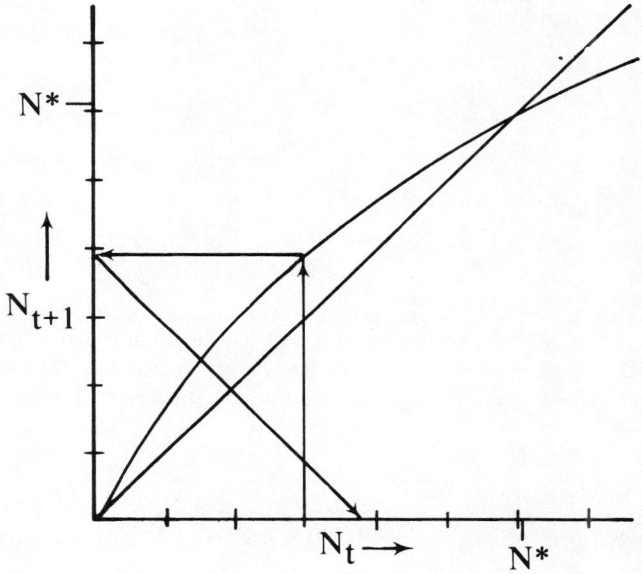

Figure 10.2. A stock-recruitment or reproductive curve. The arrows indicate one cycle of use of the diagram. N^* is the equilibrium point.

the number of the next older age class at time $(t + 1)$. The number of each age times the age-specific reproductive rate produces the contribution of that age class to the newborn, and the sum of these over all ages will be the number in the youngest age class at time $(t + 1)$. The process is repeated to calculate the changes for the following interval, using the same set of rates, or rates modified perhaps by density-dependence; Frank (1960) provides an example of the latter. The method is usually applied to the age composition and specific rates for females only to simplify the situation. Example 10.2 illustrates the procedure.

These operations are neatly systematized by the notation and methods of matrix algebra. Leslie (1945) proposed and developed this procedure, and subsequently he (1948) and others have extended the use of matrix algebra in population mathematics. Because of his contributions the matrix described below is often called a "Leslie matrix"; another term used is "projection matrix." In it, the age-specific reproductive rates and age-specific survival rates (applying to females only) are arranged in a square matrix. This is illustrated by the following for a

Ex. 10.2. Predicting population growth from age-specific survival and reproductive rates. Daphnia were reared in a culture medium; the population consisted only of females which reproduced by parthenogenesis. Below densities of 500 individuals per liter of the medium, the population had the following age-specific survival and reproductive rates.

Age in Days	s_a	b_a
0	.9	0
8	.9	6
16	.5	8
24	0	1

A culture was started with 25 individuals having the age distribution shown in the second column of the accompanying table (N_a is the number of each age). This table details the calculations for projecting the population through two eight-day periods. Predicted results are rounded to the nearest integer.

Age in	$t = 0$			$t = 8$			$t = 16$	
Days	N_a	$N_a \times b_a = B$	$N_a \times s_a$	N_a	$N_a \times b_a = B$	$N_a \times s_a$	N_a	$N_a \times b_a = B$
0	9	9 × 0 = 0		98	98 × 0 = 0		107	107 × 0 = 0
8	8	8 × 6 = 48	9 × .9 =	8	8 × 6 = 48	98 × .9 =	88	88 × 6 = 528
16	6	6 × 8 = 48	8 × .9 =	7	7 × 8 = 56	8 × .9 =	7	7 × 8 = 56
24	2	2 × 1 = 2	6 × .5 =	3	3 × 1 = 3	7 × .5 =	4	4 × 1 = 4
$N.$ =	25	$B.$ = 98	$N.$ = 116		$B.$ = 107	$N.$ = 206		$B.$ = 588

population with only four age classes (meaning $s_3 = 0$).

$$\begin{bmatrix} b_0 & b_1 & b_2 & b_3 \\ s_0 & 0 & 0 & 0 \\ 0 & s_1 & 0 & 0 \\ 0 & 0 & s_2 & 0 \end{bmatrix} \tag{10.5}$$

The age composition of the (female) population at time t is presented in a matrix consisting of only one column, that is, a column vector:

$$\begin{bmatrix} N_{0t} \\ N_{1t} \\ N_{2t} \\ N_{3t} \end{bmatrix} \tag{10.6}$$

To explain the steps in multiplication of a matrix by a vector, the following symbols and definitions are used:

A_{ij} : one element of a matrix, i indicating the horizontal row and j the vertical column in each of which the element is positioned (rows are numbered from top downward, columns from left to right).

B_i : one element of a column vector, i indicating its position in the column.

C_i : one element of the product vector, the result of multiplying matrix A by vector B.

The matrix is square, both i and j range from 1 to m. Both column vectors are of the same order, i ranging from 1 to m. Each element of the product vector

$$C_i = \Sigma_{k=1}^{m} A_{ik}B_k = A_{i1}B_1 + A_{i2}B_2 + \ldots + A_{im}B_m.$$

Applying these rules to the multiplication of the projection matrix (10.5) and the column vector (10.6):

$$\begin{bmatrix} b_0 & b_1 & b_2 & b_3 \\ s_0 & 0 & 0 & 0 \\ 0 & s_1 & 0 & 0 \\ 0 & 0 & s_2 & 0 \end{bmatrix} \times \begin{bmatrix} N_{0t} \\ N_{1t} \\ N_{2t} \\ N_{3t} \end{bmatrix} \rightarrow \begin{bmatrix} b_0N_{0t} + b_1N_{1t} + b_2N_{2t} + b_3N_{3t} \\ s_0N_{0t} + 0 + 0 + 0 \\ 0 + s_1N_{1t} + 0 + 0 \\ 0 + 0 + s_2N_{2t} + 0 \end{bmatrix} = \begin{bmatrix} N_{0,t+1} \\ N_{1,t+1} \\ N_{2,t+1} \\ N_{3,t+1} \end{bmatrix} \qquad (10.7)$$

The product vector is the age composition for time (t + 1), and it can be used with the projection matrix to predict the population for time (t + 2), and so on. The same projection matrix may be used each time, or it may be modified as the population density or other conditions change. An example of doing the latter is reported by Pennycuick *et al.* (1968). Example 10.3 uses matrix notation for the first step of example 10.2.

Ex. 10.3. A Leslie or projection matrix. The parameters of example 10.2 for $t = 0$ would be set up as below and the results would be the same as reported in that example.

$$\begin{bmatrix} 0 & 6 & 8 & 1 \\ .9 & 0 & 0 & 0 \\ 0 & .9 & 0 & 0 \\ 0 & 0 & .5 & 0 \end{bmatrix} \times \begin{bmatrix} 9 \\ 8 \\ 6 \\ 2 \end{bmatrix} = \begin{bmatrix} 98 \\ 8 \\ 7 \\ 3 \end{bmatrix}$$

The matrix algebra calculations just illustrated are no different from those described in the first paragraph of this section. This raises the question, Why use matrix algebra? There are at least three reasons: it does a tidy job, especially when the calculations are done repeatedly; matrix algebra is easy to use in a computer; and there are other, more difficult problems in population mathematics that can best be attacked by matrix algebra. Short introductions to matrix algebra for ecologists are presented by Poole (1974, sec. 1–5) and Williamson (1972, pp. 21–26, 155–59); both authors and also Usher (1972) describe some of the applications of matrix algebra to populations. Pielou (1969) presents a more mathematical treatment and includes the calculation of the stable age distribution from the projection matrix. Keyfitz (1968) describes many uses of the projection matrix in the study of human populations; his treatment requires a thorough understanding of matrix algebra. Keyfitz and Flieger (1971, chap. 8) include a computer program for the matrix population model.

10.4 STOCHASTIC MODELS

The models so far discussed have been deterministic, because for each set of parameters and initial conditions only one result is predicted for each future time. "Stochastic" models introduce some element of chance or some random process that, independently of the population characteristics, affects the results. One kind of stochastic model is illustrated in table 10.1 and figure 10.3. The table sets forth the probabilities of three different growth factors, ratios of N_{t+1} to N_t. In

Table 10.1. Values of the Growth Ratio and the Probabilities of Each for a Simple Stochastic Model of Population Growth

$g_t = \frac{N_{t+1}}{N_t}$	$P(g_t)$
1.5	0.6
1.0	0.3
0.67	0.1

$$E(g_t) = 1.27$$

the figure the centers of the circles are the possible values of N_t and the area of each circle is proportional to the probability of that value. The possible routes to each N_t are shown, and the probability is the sum of the probabilities of arriving at that point by all possible routes.

The heavy line is both the average and the expected growth, average in that it is the mean of the possible values of N_t weighted by the probability of each, expected in that it is the geometric growth produced by the expected (mean) value of g_t = 1.27.

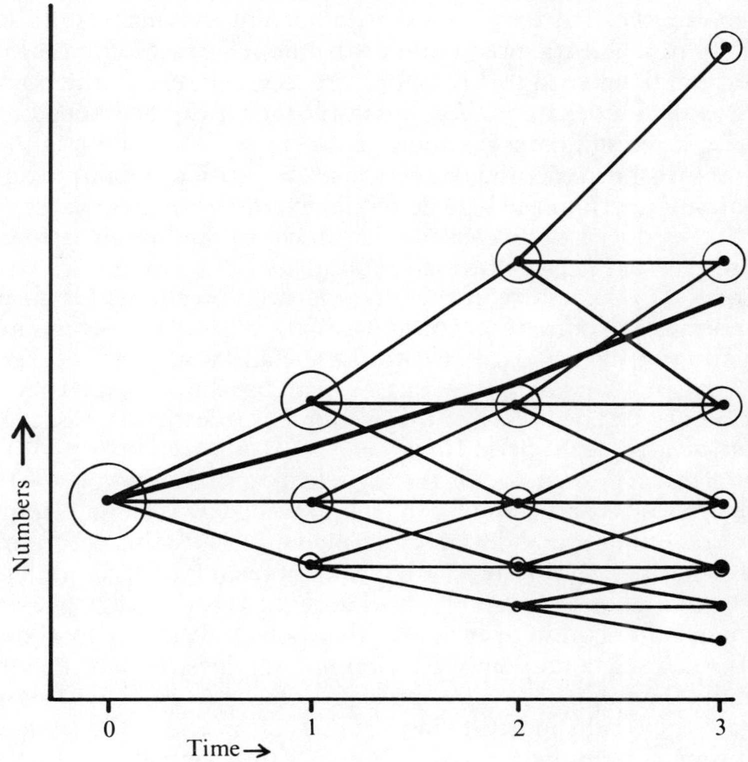

Figure 10.3. A stochastic model of population growth. See text for explanation.

The figure illustrates these points about stochastic models. The primary one is that there are several possible outcomes, each with its own probability of occurring. Another is that although in this case the probabilities are on the side of population increase, there is a finite probability of the population decreasing even to the point of extinction. In this example, the average or expected N_t are predictable from a deterministic model; this is not true, however, for all stochastic models.

The above example of a stochastic model of population growth illustrates the use of a Markov chain, where the probabilities used in each transition lead to a calculation of the probability of each outcome. Another way of introducing chance or random events into a model is the Monte Carlo method. In this a random variable is introduced into some part of the model, the resulting outcome is calculated, and the process is repeated several times with different values of the random variable. The frequency distribution of the final results determines the probabilistic prediction of the model. The Monte Carlo method can be likened to a sampling process, wherein all the possible results of the model are randomly sampled, or it can be likened to analyzing the results of a gambling game.

The Monte Carlo method can be illustrated with an example of population growth in the logistic model in which N_0 and r_m are fixed but K, the carrying capacity, is subject to random variation. The assumptions that need to be made about the latter are (1) the mean value of K; (2) the nature of the random variation about this mean (e.g., normal, Poisson, or uniform distribution); and (3) the amount of variation or the standard deviation to be allowed. For this example, $N_0 = 2$, $r_m = .4$, and $\overline{K} = 40$ with a normal distribution and a standard deviation of 10. The object of this example is to estimate N_8, the population after eight periods.

The difference equation for the logistic model, equation (10.4), will be used. The first step of one trial proceeds by selecting from a table of random normal deviates a chance value. Let this be, for example, +.5; then $K_0 = \overline{K} + .5S = 45$. Substituting into equation (10.4), $K = 45$, $N_0 = 2$, and $e^r = e^{.4} = 1.49$; the result is $N_1 = 2.92$. The calculation is reiterated with each resulting value of N_t for a total of eight times, using a randomly selected normal deviate each time to determine the value of K. The entire process was repeated 20 times to obtain the results illustrated in figure 10.4, in which the frequency distribution of 20 values of N_8 is shown in the graph at the right of the figure.

Any one or more variables in a model can be made a random variable in the Monte Carlo method. For example, equation (10.4) could have been used with a constant r_m and a constant K and with a random element added to the calculated N_{t+1}. The resulting frequency distribution of N_8 would not have been the same as in the example given. The choice between models must be made on the ultimate basis of which is most realistic, paralleling the results observed in nature.

Stochastic models, because they allow for random processes which may be introduced into the model in many different ways, are more realistic than are deterministic models. Despite this, they have been little used in population modeling because with only slight complexity they become intractable. Chapman (1967) in a review of stochastic models in animal population ecology concluded with this statement: "That the models considered are much too simple is fully agreed—yet it is also true that these simple models are often insolvable by present analytic technics." The use of numerical technics to assess models by computer simulation is discussed in the next section.

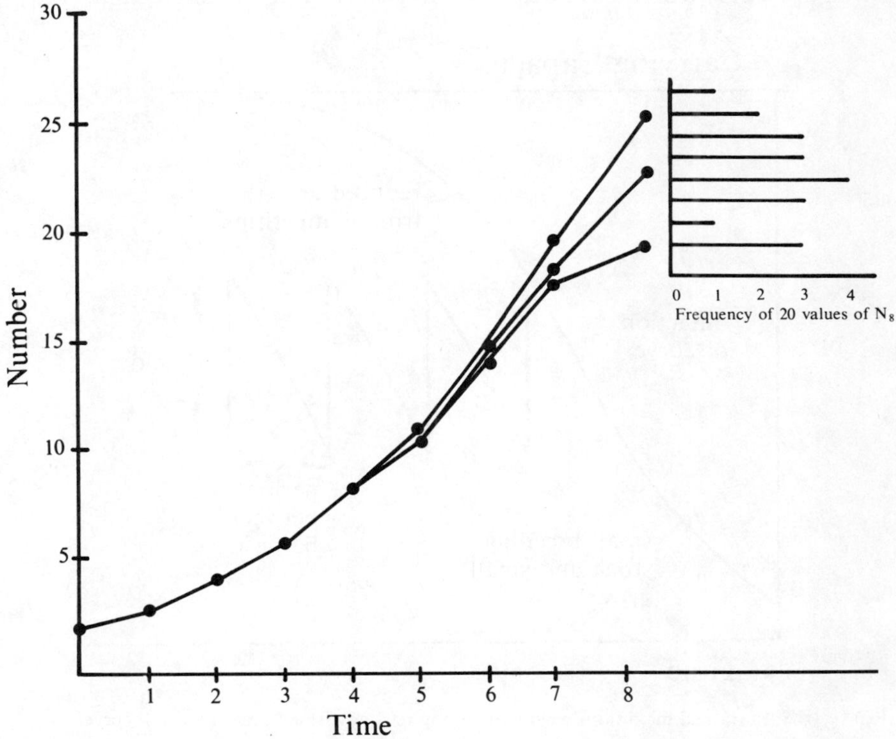

Figure 10.4. Stochastic growth of a population. A Monte Carlo method was used with the logistic difference equation to predict 20 values of N_8; see text for explanation. Only three of the calculated growth curves are plotted in full.

Bailey (1964) has written a text on stochastic processes applicable to the study of biological populations. Poole (1974) describes some stochastic models of population growth differing from the examples used in this section. Bartlett's (1973) paper is a brief exposition of stochastic models using matrix notation.

10.5 OPTIMUM YIELD AND COMPUTER SIMULATION

The goal of most management of biological resources is to obtain an optimum yield or harvest; usually this is the maximum harvest which can be sustained indefinitely. For some harvests, such as commercial fisheries, the goal is to obtain the maximum sustainable annual harvest of weight of fish; for others, as in quail management, the crop is measured in numbers. The basic problem can be illustrated with the aid of figure 10.5, which assumes some population possessing a sigmoid

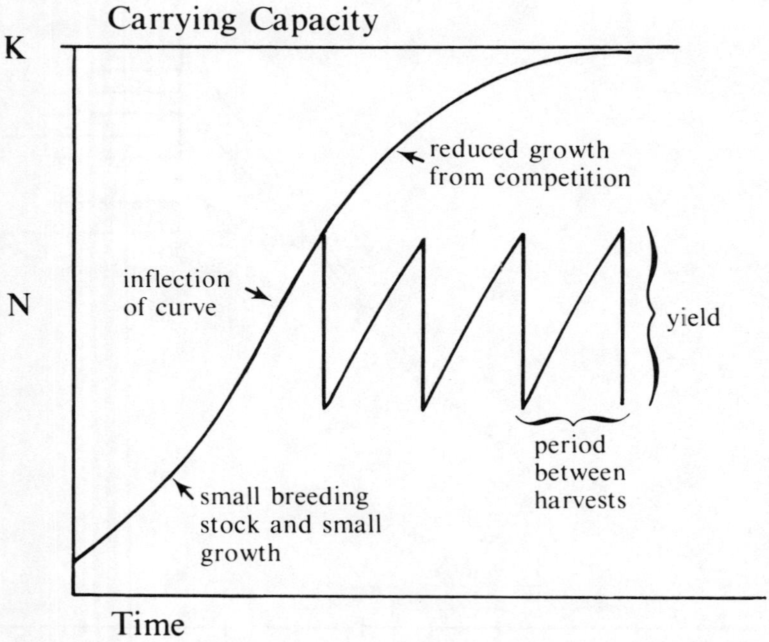

Figure 10.5. Sustained maximum yield from a population with a sigmoid growth curve.

growth curve and, for the moment a stable carrying capacity. The maximum sustained yield could be obtained by continually harvesting the population so as to maintain the numbers at the point where the curve is the steepest; below this point the growth, and therefore the possible harvest, is less because the breeding stock is smaller, and above this point the growth is less because competition is increasing as the numbers approach the carrying capacity. A continual harvest, however, is the exception rather than the rule, and most resources are cropped at intervals, as is illustrated in the figure. This means that there are three interdependent management decisions: the frequency of cropping, the size of the crop, and the size of the breeding stock or population base to be left to produce the next harvest.

The yield could be further increased by improving the carrying capacity of the environment. It can be seen intuitively that if K were increased, the inflection point of the curve should move up and become steeper, thus increasing the potential yield.

The above description of the optimum yield problem is oversimplified in that it neglects uncontrollable changes in the environment such as those produced by weather, and it overlooks the fact that almost any form of harvesting will change the sex and age composition of the population and therefore the potential population growth rate. These factors may be included in the following general way. Let N_{it} be the number of any class (age, sex, etc.) in the population at time t, and H_{it} be the number removed from that class. Then ΣH_{it} is the total harvest, and $N_{it} - H_{it} = S_{it}$ is the number of each class surviving that harvest. The number of each class available to the next harvest at $(t + 1)$ will be determined by the population changes occurring until then, and will be dependent upon the number of survivors in each class and the environmental conditions. In very general terms, $N_{i(t+1)} = f(S_{1t}, S_{2t}, \ldots S_{kt}; E_t)$ where E_t summarizes the environmental conditions between t and $(t + 1)$. Included in this function will be such population characteristics as age-specific fecundity, age-specific survival, competition within and between individuals of each class, and the effects of the predicted environmental conditions on each of the population processes. The next harvest will then be $\Sigma H_{i(t+1)}$. The optimum yield problem then becomes one of choosing the H_i so as to maximize the ΣH_i in this harvest and in each succeeding harvest. If the goal is to maximize the harvest weight, not just the numbers, the problem becomes even more complex because, among other things, the gain in weight of each individual surviving from one harvest to the next will need to be predicted.

To include the effects of all the factors in the solution of the above problem is obviously not a simple calculation. Fortunately, computers enable us to attack problems of this complexity.

The characteristics of computers which make this possible are their speed of operation, especially useful in cases requiring iterative calculations, their memories in which can be stored the results of intermediate calculations and such items as logarithms and random numbers, and their tremendous flexibility in operation. A computer can be directed to perform many kinds of mathematical operations—calculus, transformations, matrix algebra—in any sequence by a computer language such as Fortran. A program so complex that its writer cannot comprehend the whole in its entirety can be built in steps, each step being some component of the problem. The components can be tested separately for performance, and then all combined into the complete model.

One use of computer models is to simulate population changes for predictive purposes. Another is to test various management schemes. The latter may be nothing more than a trial and error process, changing one or more of the variables until the model's output equals the management goals. Once the model is constructed, experiments of this sort are easy because of the speed of computer calculations.

Watt (1968) describes the use of computers in natural resources management, including procedures in the construction of models and giving various examples. Paulik and Greenough (1966) review models used in fishery management and develop one of a salmon fishery. Henny *et al.* (1970) describe a model intended to predict among other things the amount of allowable harvest of game while maintaining a stable population. Beddington (1974) applies a matrix model like that described in section 10.3 to the general problem of determining an optimal harvest. Huang *et al.* (1976) present mathematical proofs for the solution of optimal yield from fisheries for both deterministic and stochastic models. Other citations may be found in a bibliography on quantitative ecology (Schultz *et al.*, 1976, pp. 103–28).

A very important point needs to be made in concluding this subject, and that is that a computer model, no matter how complex, is no substitute for information on population characteristics. In fact, computers are "data hungry." Some data are needed as population parameters to serve as inputs or controlling variables. The output of even the most sophisticated model will be no more reliable that the inputs are accurate; as computer programmers say, "Garbage in means garbage out." Other information is needed for validating the model, done by comparing the model's output with observations of real populations. All of the many characteristics described in the preceding chapters may be significant in computer simulations of populations, but the usefulness of the simulations will depend upon how well the characteristics are understood and how accurately they are measured. The same may be said, of course, for any study of populations, whether made for theoretical or practical purposes.

Literature Cited

Aldrich, J. W., and C. S. Robbins. 1970. Changing abundance of migratory birds in North America. *Smithsonian Contr. Zool.* 26:17–26.

Allee, W. C., A. E. Emerson, O. Park, T. Park, and K. P. Schmidt. 1949. *Principles of animal ecology.* Saunders, Philadelphia. 837 pp.

Armitage, P. 1971. *Statistical methods in medical research.* John Wiley, New York. xv + 504 p.

Arnold, D. E. 1966. *Marking fish with dyes and other chemicals.* Bur. Sport Fish. and Wildl., Tech. Paper 10.44 pp.

Bagenal, T. B. 1967. A short review of fish fecundity. In *Biological basis of freshwater fish production*, ed. S. D. Gerking, pp. 89–111. Blackwell Sci. Publ., Oxford.

Bailey, N. T. J. 1951. On estimating the size of mobile populations from recapture data. *Biometrika* 38: 293–306.

——. 1952. Improvements in the interpretation of recapture data. *J. Anim. Ecol.* 21: 120–27.

——. 1964. *The elements of stochastic processes; with applications to the natural sciences.* John Wiley, New York. xi + 249 pp.

Barclay, G. W. 1958. *Techniques of population analysis.* John Wiley & Sons, New York. xiii + 311 pp.

Bartlett, M. S. 1973. Equations and models of population change. In *Mathematical theory of the dynamics of biological populations*, ed. M. S. Bartlett and R. W. Hiorns, pp. 5–21. Academic Press, New York.

——, and R. W. Hiorns, eds. 1973. *The mathematical theory of the dynamics of biological populations.* Academic Press, New York. xii + 348 pp.

Beddington, J. R. 1974. Age structure, sex ratio, and population density in the harvesting of natural populations. *J. Appl. Ecol.* 11: 915–24.

Bergerud, A. 1968. Numbers and density. In *A practical guide to the study of productivity of large herbivores*, ed. F. B. Golley and H. K. Buechner, pp. 21–42. Blackwell Sci. Publ., Oxford.

Berthold, P. 1976. Methoden der Bestandserfassung in der Ornithologie: Ubersicht and kritische Betrachtung. *J. Ornithol.* 117: 1–69.

Best, L. B. 1975. Interpretational errors in the "mapping method" as a census technique. *Auk* 92: 452–60.

Beverton, R. J. H., and S. J. Holt. 1957. *On the dynamics of exploited fish populations.* Her Majesty's Stationery Office, London. 533 pp.

Birch, L. C. 1948. The intrinsic rate of natural increase of an insect population. *J. Anim. Ecol.* 17: 15–26.

Blackith, R. E. 1958. Nearest neighbor distance measurement for the estimation of animal populations. *Ecology* 39: 147–50.

Botkin, D. B., and R. S. Miller. 1974. Mortality rates and survival of birds. *Amer. Naturalist* 108: 181–92.

Campbell, R. W. 1969. Studies in gypsy moth population dynamics. In *Forest Insect Population Dynamics,* pp. 29–34. USDA Forest Serv. Res. Papers NE-125.

Caswell, H. 1972. On instantaneous and finite birth rates. *Limnol. and Oceanogr.* 17: 787–91.

Caughley, G. 1966. Mortality patterns in mammals. *Ecology* 47: 906–18.

——. 1974. Interpretation of age ratios. *J. Wildl. Mgmt.* 38: 557–62.

——, and L. C. Birch. 1971. Rate of increase. *J. Wildl. Mgmt.* 35: 658–63.

Chapman, D. G. 1955. Population estimation based on change of composition caused by selective removal. *Biometrika* 42: 279–90.

——. 1967. Stochastic models in animal population ecology. In *Proc. 5th Berkely Symp. Math. Stat. and Probability,* ed. L. M. LeCam and J. Neyman, pp. 147–62. Univ. California Press, Berkeley.

Chapman, D. G., and W. S. Overton. 1966. Estimating and testing differences between population levels by the Schnabel estimation method. *J. Wildl. Mgmt.* 30: 173–80.

Chapman, D. G., and D. S. Robson. 1960. The analysis of a catch curve. *Biometrics* 16: 354–68.

Coale, A. J. 1972. *The growth and structure of human populations.* Princeton Univ. Press, Princeton, N. J. 244 pp.

Cochran, W. G. 1963. *Sampling techniques*, 2d ed. Wiley, New York. 344 pp.

Cole, L. C. 1954. The population consequences of life history phenomena. *Quart. Rev. Biol.*, 29: 103–37.

Conover, W. J. 1971. *Practical nonparametric statistics.* Wiley, New York. 472 pp.

Conway, G. R. 1973. Experience in insect pest modelling: a review of models, uses and future directions. In *Insects: studies in population management*, ed. P. W. Geir *et al.*, pp. 103–30. *Ecol. Soc. Australia Mem.* 1, Canberra.

Cormack, R. M. 1968. The statistics of capture-recapture methods. *Oceanog. Mar. Biol. Ann. Rev.* 6: 455–506.

——. 1970. [Appendix to Construction of life tables from the recovery of dead animals, by R. A. Fordham.] *J. Anim. Ecol.* 39: 24–27.

Costello, W. G., and H. M. Taylor. 1971. Deterministic population growth models. *Amer. Math Mo.* 78: 841–55.

Cushing, D. H. 1968. *Fisheries biology: a study in population dynamics.* Univ. Wisconsin Press, Madison. 200 pp.

Darling, F. F. 1951. The ecological approach to the social sciences. *Amer. Sci.* 39: 244–54.

Deevey, E. S., Jr. 1947. Life tables for natural populations of animals. *Quart. Rev. Biol.* 22: 283–314.

Dempster, J. P. 1975. *Animal population ecology.* Academic Press, New York. x + 155 p.

Dublin, L. I., and A. J. Lotka. 1936. *Length of life: a study of the life table.* Ronald Press, New York. 422 pp.

Eberhardt, L. L. 1969. Population estimates from recapture frequencies. *J. Wildl. Mgmt.* 33: 28–39.

——. 1971. Population analysis. In *Wildlife management techniques,* ed. R. H. Giles, pp. 457–95. Wildlife Soc., Washington, D. C.

——. 1972. Some problems in estimating survival from banding data. In *Population ecology of migratory birds,* pp. 153–71. Wildl. Res. Rept. 2, Bur. Sport Fish. and Wildl., U.S.D. 1.

——. 1973. *Line transects for population studies.* Pacific N. W. Lab. Ann. Rept. 1973. Part 2, Ecol. Studies. pp. 3–4.

Ebert, T. A. 1973. Estimating growth and mortality rates from size data. *Oecologia* 11: 281–98.

Edmonson, W. T., and G. G. Winberg, eds. 1971. *A manual on methods for the assessment of secondary productivity in fresh waters.* Blackwell Sci. Publ., Oxford. 325 pp.

Elliott, J. M. 1971. *Some methods for the statistical analysis of samples of benthic invertebrates.* Freshwater Biol. Assoc. Sci. Publ. 25. 144 pp.

Emlen, J. T. 1940. Sex and age ratios in the survival of the California Quail. *J. Wildl. Mgmt.* 4: 92–99.

——. 1971. Population densities of birds derived from transect counts. *Auk* 88: 323–41.

Farner, D. S. 1955. Bird banding in the study of population dynamics. In *Recent studies in avian biology,* ed. A. Wolfson, pp. 397–499. Univ. Illinois Press, Urbana.

Farris, J. S. 1968. Significance testing and confidence intervals for fixed mortality rates. *Ecology* 49: 994–96.

Finney, D. J. 1962. *Probit analysis.* 2d ed. Cambridge Univ. Press, Cambridge. 318 pp.

Frank, P. W. 1960. Prediction of population growth form in *Daphnia pulex* cultures. *Amer. Naturalist* 94: 357–72.

French, N. R., C. D. Jorgensen, M. H. Smith, and B. G. Maza. 1971. *Comparison of some IBP population estimate methods for small mammals.* USNC/IBP Special Report July 1971. 25 pp.

Gadgil, M., and W. H. Bossert. 1970. Life history consequences of natural selection. *Amer. Naturalist* 104: 1–24.

Gallopin, G. C. 1971. A generalized model of a resource-population system. *Oecologia* 7: 382–432.

Geis, A. D., and R. D. Taber. 1963. Measuring hunting and other mortality. In *Wildlife investigational techniques,* ed. H. S. Mosby, pp. 284–98. 2d ed. Wildlife Soc., Washington, D. C.

Gilbert, R. O. 1973. Approximations of the bias in the Jolly-Seber capture-recapture model. *Biometrics* 29: 501–26.

Giles, R. H., Jr. 1969. Topographic model of population stability. *J. Wildl. Mgmt.* 33: 1042–45.

——, ed. 1971. *Wildlife management techniques,* 3d ed. Wildlife Soc., Washington, D. C. vii + 633 pp.

Goodall, D. W. 1970. Statistical plant ecology. *Ann. Rev. Ecol. System.* 1: 99–124.

Goodman, L. A. 1971. On the sensitivity of the intrinsic growth rate to change in the age-specific birth and death rates. *Theor. Popul. Biol.* 2: 339–54.

Greig-Smith, P. 1964. *Quantitative plant ecology*, 2d ed. Butterworth, Washington. 268 pp.

Haldane, J. B. S. 1955. The calculation of mortality rates from ringing data. *Acta XI Congr. Intern. Ornith.*: 454–58.

Hall, G. A. 1964. Breeding-bird censuses—why and how. *Audubon Field Notes* 18: 413–16.

Hanson, W. R. 1963. Calculation of productivity, survival, and abundance of selected vertebrates from the sex and age ratios. *Wildl. Monogr.* 9. 60 pp.

Harcourt, D. G. 1969. The development and use of life tables in the study of natural insect populations. *Ann. Rev. Entomol.* 14: 175–96.

——. 1971. Population dynamics of *Leptinotarsa decemlineata* (Say) in eastern Ontario. *Canad. Entomol.* 103: 1049–61.

Hayne, Don. W. 1949. An examination of the strip census method for estimating animal populations. *J. Wildl. Mgmt.* 13: 145–57.

Henny, C. J., W. S. Overton, and H. M. Wight. 1970. Determining parameters for populations by using structural models. *J. Wildl. Mgmt.* 34: 690–702.

Hickey, J. J. 1952. *Survival studies of banded birds.* U. S. Fish and Wildl. Serv. Spec. Sci. Rpt., Wildl. No. 15. 177 pp.

Hjort, J. 1926. Fluctuations in the year classes of important food fishes. *J. du Conseil Perm. Intern. pour L'Exploration de la Mer*, 1: 1–38.

Huang, C. C., I. B. Vertinsky, and N. J. Wilimovsky. 1976. Optimal controls for a single species fishery and the economic value of research. *J. Fish. Res. Board Canada* 33: 793–809.

Itô, Yosiaki. 1972. On the methods for determining density-dependence by means of regression. *Oecologia* 10: 347–72.

Jeffers, J. N. R., ed. 1972. *Mathematical models in ecology.* Blackwell Sci. Publ., Oxford.

Jolly, G. M. 1965. Explicit estimates from capture-recapture data with both death and immigration—stochastic model. *Biometrika* 52: 225–47.

Kadlec, J. A., and W. H. Drury. 1968. Structure of the New England Herring Gull population. *Ecology* 49: 644–76.

Kelker, G. H., and W. R. Hanson. 1964. Simplifying the calculation of differential survival of age-classes. *J. Wildl. Mgmt.* 28: 411.

Kendeigh, S. C. 1944. Measurement of bird populations. *Ecol. Monogr.* 14: 67–106.

Kershaw, K. A. 1973. *Quantitative and dynamic ecology,* 2d ed. Edward Arnold, London. x + 308 pp.

Keyfitz, Nathan. 1968. *Introduction to the mathematics of populations.* Addison-Wesley, Reading, Mass. xiv + 450 pp.

——. 1971. Changes in birth and death rates and their demographic effects. In *Rapid population growth,* ed. Roger Revelle *et al.*, pp. 639–80. Johns Hopkins Press, Baltimore.

——, and W. Flieger. 1971. *Population: facts and methods of demography.* W. H. Freeman Co., San Francisco. 614 pp.

Lagler, K. F. 1971. Capture, sampling and examination of fishes. In *Methods for assessment of fish production in fresh waters*, ed. W. E. Ricker, pp. 7–44. Blackwell Sci. Publ., Oxford.

Lamotte, M., ed. 1969. *Problems d'ecologie: l'echantillonnage des peuplements animaux des milieux terrestres.* Masson et Cie., Paris. 303 pp.

Laughlin, R. 1965. Capacity for increase: a useful population statistic. *J. Anim. Ecol.* 34: 77–91.

Leslie, P. H. 1945. On the use of matrices in certain population mathematics. *Biometrika* 33: 183–212.

——. 1948. Some further notes on the use of matrices in population mathematics. *Biometrika* 35: 213–45.

——. 1966. The intrinsic rate of increase and the overlap of successive generations in a population of Guillemots (*Uria aalge* Pont.) *J. Anim. Ecol.* 35: 291–301.

Lotka, A. J. 1922. The stability of the normal age distribution. *Proc. Nat. Acad. Sci.* 8: 339–45.

Ludwig, J. P. 1967. Band loss—its effect on banding data and apparent survivorship in the ring-billed gull population of the Great Lakes. *Bird-Banding* 38: 309–23.

Macfadyen, A. 1962. Soil arthropod sampling. *Adv. Ecol. Res.* 1: 1–34.

Manly, B. F. J. 1971. Simulation study of Jolly's method for analyzing capture-recapture data. *Biometrics* 27: 415–24.

May, R. M. 1974. *Stability and complexity in model ecosystems.* 2d ed. Princeton Univ. Press, Princeton, N. J. ix + 265 pp.

——, ed. 1976. *Theoretical ecology: principles and applications.* W. B. Saunders Co., Philadelphia. viii + 317 pp.

May, R. M., G. R. Conway, M. P. Hassell, and T. R. E. Southwood. 1974. Time delays, density dependence and single-species populations. *J. Anim. Ecol.* 43: 747–70.

Mayfield, H. F. 1975. Suggestions for calculating nest success. *Wilson Bul.* 87: 456–66.

Maynard Smith, J. 1974. *Models in ecology.* Cambridge Univ. Press, Cambridge. xii + 146 pp.

McKnight, M. E. 1971. Natural mortality of the western spruce budworm *Choritoneura occidentalis* in Colorado. U. S. D. A. Forest Serv. Res. Paper RM-81. 12 pp.

Morris, R. F. 1965. Contemporaneous mortality factors in population dynamics. *Canad. Entomol.* 97: 1173–84.

——, and Miller, C. A. 1954. The development of life tables for the spruce budworm. *Canad. J. Zool.* 32: 283–301.

Mueller-Dumbois, D., and H. Ellenberg. 1974. *Aims and methods of vegetation ecology.* Wiley, New York. 547 p.

Mukerji, M. K. 1971. Major factors in survival of the immature stages of *Hylemya brassicae* (Diptera: Anthomyiidae) on cabbage. *Canad. Entomol.* 103: 717–28.

Murie, A. 1944. *The wolves of Mount McKinley.* U. S. Nat. Park Serv., Fauna series no. 5. xx + 238 pp.

Murphy, P. W. 1962. Extraction methods for soil animals. In *Progress in soil zoology,* ed. P. W. Murphy, pp. 75–114, 115–55. Butterworth, London.

Myers, K. 1971. The rabbit in Australia. In *Dynamics of populations,* ed. P. J. den Boer and G. R. Gradwell, pp. 478–506. Wageningen, Netherlands.

Overton, W. S. 1971. Estimating the numbers of animals in wildlife populations. In *Wildlife management techniques,* ed. R. H. Giles, Jr., pp. 403–55 3d ed. Wildlife Society, Washington, D. C.

Paloheimo, J. E. 1974. Calculation of instantaneous birth rate. *Limnol. and Oceanogr.* 19: 692–94.

Patil, G. P., E. C. Pielou, and W. E. Waters, eds. 1971. *Statistical ecology.* Penn. State Univ. Press. 3 vols.

Paulik, G. J., and J. W. Greenough, Jr. 1966. Management analysis for a salmon resource system. In *Systems analysis in ecology,* ed. K. E. F. Watt, pp. 215–52. Academic Press, New York.

——, and D. S. Robson. 1969. Statistical calculations for change-in-ratio estimators of population parameters. *J. Wildl. Mgmt.* 33: 1–27.

Pennycuick, C. J., R. M. Compton, and L. Beckingham. 1968. A computer model for simulating the growth of a population, or of two interacting populations. *J. Theoret. Biol.* 8: 316–29.

Petrides, G. A. 1949. Viewpoints on the analysis of open season sex and age ratios. *Trans. N. Amer. Wildl. Conf.* 14: 391–410.

Petrusewicz, K., and A. Macfadyen. 1970. Productivity of terrestrial animals: principles and methods. *Intern. Biol. Progr. Hdbk. No. 13*. Davis, Philadelphia. 202 pp.

Phillips, B. F., and N. A. Campbell. 1970. Comparisons of methods of estimating population size using data on the whelk. *J. Anim. Ecol.* 39: 753–59.

Pielou, E. C. 1969. *An introduction to mathematical ecology.* Wiley, New York. vii + 286 pp.

Podoler, H., and D. Rogers. 1975. A new method for the identification of key factors from life table data. *J. Anim. Ecol.* 44: 85–114.

Pollock, K. H. 1975. A *K*-sample tag-recapture model allowing for unequal survival and catchability. *Biometrika* 62: 577–83.

Poole, R. W. 1972. An autoregressive model of population density changes in an experimental population of *Daphnia magna. Oecologia* 10: 205–21.

——. 1974. *An introduction to quantitative ecology.* McGraw-Hill, New York. 480 pp.

Pucek, Z., and V. P. W. Lowe. 1975. Age criteria in small mammals. In *Small mammals: their productivity and population dynamics,* eds. F. B. Golley, K. Petrusewicz, and L. Ryszkowski, pp. 55–72. Cambridge Univ. Press, Cambridge.

Reeves, J. E. 1972. The distribution of the maximum likelihood estimator of the parameter in the first-order autoregressive series. *Biometrika* 59: 387–94.

Regier, H. A., and D. S. Robson. 1967. Estimating population number and mortality rates. In *The biological basis of freshwater fish production*, ed. S. D. Gerking, pp. 31–66. Blackwell Sci. Publ., Oxford.

Ricker, W. E. 1954. Stock and recruitment. *J. Fish. Res. Bd. Canada*, 11: 559–623.

——. 1975. Computation and interpretation of biological statistics of fish populations. *Fish. Res. Bd. Canada Bul.* 191. xvii + 382 pp.

Robbins, C. S., and W. T. van Velzen. 1970. Progress report on the North American breeding bird survey. In *Bird census work and environmental monitoring*, ed. S. Svensson, pp. 22–30. Ecol. Res. Comm. Bul. 9, Swedish Res. Council, Stockholm.

Robson, D. S., and D. G. Chapman. 1961. Catch curves and mortality rates. *Amer. Fish. Soc. Trans.* 90: 181–89.

——, and W. A. Flick. 1965. A nonparametric statistical method for culling recruits from a mark-recapture experiment. *Biometrics* 21: 936–47.

——, and H. A. Regier. 1966. Estimates of tag loss from recoveries of fish tagged and permanently marked. *Amer. Fish. Soc. Trans.* 95: 56–59.

Roff, D. A. 1973a. On the accuracy of some mark-recapture estimators. *Oecologia* 12: 15–34.

——. 1973b. Examination of some statistical tests used in the analysis of mark-recapture data. *Oecologia* 12: 35–54.

Schnabel, Z. E. 1938. The estimation of the total fish population in a lake. *Amer. Math. Monthly* 45: 348–52.

Schultz, V., L. L. Eberhardt, J. M. Thomas, and M. I. Cochran. 1976. *A bibliography of quantitative ecology*. Dowden, Hutchinson and Ross, Stroudsburg, Pa. vii + 361 pp.

——, and R. J. Muncy. 1957. An analysis of variance applicable to transect population data. *J. Wildl. Mgmt.* 21: 274–78.

Schumacher, F. X., and R. W. Eschmeyer. 1943. The estimate of fish population in lakes or ponds. *J. Tenn. Acad. Sci.* 18: 228–49.

Seber, G. A. F. 1965. A note on the multiple-recapture census. *Biometrika* 52: 249–59.

——. 1970. Estimating time-specific survival and reporting rates for adult birds from band returns. *Biometrika* 57: 313–18.

——. 1972. Estimating survival rates from bird-band returns. *J. Wildl. Mgmt.* 36: 405–13.

——. 1973. *The estimation of animal abundance*. Griffin, London. xii + 506 pp.

Sen, A. R., J. Tourigny, and G. E. J. Smith. 1974. On the line transect sampling method. *Biometrics* 30: 329–40.

Siegel, S. 1956. *Nonparametric statistics for the behavioral sciences*. McGraw-Hill, New York. 312 pp.

Skellam, J. G. 1962. Estimation of animal populations by extraction processes considered from the mathematical standpoint. In *Progress in soil zoology*, ed. P. W. Murphy, pp. 26–36. Butterworth, London.

Smith, F. E. 1963. Population dynamics in *Daphnia magna* and a new model for population growth. *Ecology* 44: 651–63.

Smith, M. H., R. H. Gardner, J. B. Gentry, D. W. Kaufmam, and M. H. O'Farrell.

1975. Density estimations of small mammal populations. In *Small mammals: their productivity and population dynamics.* eds. F. B. Golley, K. Petrusewicz, and L. Ryszkowski, pp. 25–53. Cambridge Univ. Press, Cambridge.

Southwood, T. R. E. 1966. *Ecological methods.* Methuen & Co., London. 409 pp.

Spector, W. S., ed. 1956. *Handbook of biological data.* Saunders, Philadelphia, 620 pp.

Spicer, C. C. 1956. The estimation of the parameters of an exponentially declining population. *J. Hyg.* 54: 304–10.

Spiegelman, M. 1968. *Introduction to demography.* 2d ed. Harvard Univ. Press, Cambridge, Mass. xx + 514 pp.

Spinage, C. A. 1972. African ungulate life tables. *Ecology* 53: 645–52.

St. Amant, J. L. S. 1970. The detection of regulation in animal populations. *Ecology* 51: 823–28.

Stott, Brian. 1971. Marking and tagging. In *Methods for assessment of fish production in fresh waters,* ed. W. E. Ricker, pp. 82–97. Blackwell Sci. Publ., Oxford.

Streifer, W., and C. A. Istock. 1973. A critical variable formulation of population dynamics. *Ecology* 54: 392–98.

Svensson, S. 1970. *Bird census work and environmental monitoring.* Ecol. Res. Comm. Bul. 9. Swedish Res. Council. Stockholm. 52 p.

Taber, R. D. 1971. Criteria of sex and age. In *Wildlife management techniques,* ed. R. H. Giles, Jr., pp. 325–401. Wild. Soc., Washington, D. C.

Taber, R. D., and I. McT. Cowan. 1971. Capturing and marking wild animals. In *Wildlife management techniques,* ed. R. H. Giles, Jr., pp. 277–317. Wild. Soc., Washington, D. C.

Tanner, J. T. 1966. Effects of population density on growth rates of animal populations. *Ecology* 47: 733–45.

Taylor, L. R. 1961. Aggregation, variance and the mean. *Nature* 189: 723–35.

Tepper, E. E. 1967. *Statistical methods in using mark-recapture data for population estimation.* Bibl. No. 4, U. S. Dept. Interior Library. Washington, D. C.

Tesch, F. W. 1971. Age and growth. In *Methods for assessment of fish production in fresh waters*, ed. W. E. Ricker, pp. 98–130. Blackwell Sci. Publ., Oxford.

Usher, M. P. 1972. Developments in the Leslie matrix model. In *Mathematical models in ecology*, ed. J. N. R. Jeffers, pp. 29–60. Proc. 12th Symp. Brit. Ecol. Soc. Blackwell Sci. Publ., Oxford.

Varley, G. C., and G. R. Gradwell. 1960. Key factors in population studies. *J. Anim. Ecol.* 29: 399–402.

Varley, G. C., and G. R. Gradwell. 1968. Population models for the winter moth. In *Insect abundance*, ed. T. R. E. Southwood, pp. 132–42. Blackwell Sci. Publ., Oxford.

Varley, G. C., G. R. Gradwell, and M. P. Hassell. 1974. *Insect population ecology: an analytical approach.* Univ. California Press, Berkeley. x + 212 pp.

Watt, K. E. F. 1968. *Ecology and resource management.* McGraw-Hill, New York. xii + 450 pp.

Wiegert, R. G. 1974. Competition: a theory based on realistic, general equations of population growth. *Science* 185: 539–42.

Wilbur, H. M. and J. M. Landwehr. 1974. The estimation of population size with equal and unequal risks of capture. *Ecology* 55: 1339–48.

Williamson, Mark. 1972. *The analysis of biological populations.* Edward Arnold, London. 191 pp.

Yapp, W. B. 1956. The theory of line transects. *Bird Study,* 3: 93–104.

Zippin, C[alvin]. 1956. An evaluation of the removal method of estimating animal populations. *Biometrics* 12: 163–89.

——. 1958. The removal method of population estimation. *J. Wildl. Mgmt.* 22: 82–90;

Index

Index